FREE Test Taking Tips DVD Offer

To help us better serve you, we have developed a Test Taking Tips DVD that we would like to give you for FREE. **This DVD covers world-class test taking tips that you can use to be even more successful when you are taking your test.**

All that we ask is that you email us your feedback about your study guide. Please let us know what you thought about it – whether that is good, bad or indifferent.

To get your **FREE Test Taking Tips DVD**, email freedvd@studyguideteam.com with "FREE DVD" in the subject line and the following information in the body of the email:

a. The title of your study guide.

b. Your product rating on a scale of 1-5, with 5 being the highest rating.

c. Your feedback about the study guide. What did you think of it?

d. Your full name and shipping address to send your free DVD.

If you have any questions or concerns, please don't hesitate to contact us at freedvd@studyguideteam.com.

Thanks again!

8th Grade Common Core Math Workbook

Common Core Math Grade 8 Prep Book and Practice Questions [2nd Edition]

TPB Publishing

Written and edited by TPB Publishing.

ISBN 13: 9781628456653
ISBN 10: 1628456655

Table of Contents

Quick Overview

As you draw closer to taking your exam, effective preparation becomes more and more important. Thankfully, you have this study guide to help you get ready. Use this guide to help keep your studying on track and refer to it often.

This study guide contains several key sections that will help you be successful on your exam. The guide contains tips for what you should do the night before and the day of the test. Also included are test-taking tips. Knowing the right information is not always enough. Many well-prepared test takers struggle with exams. These tips will help equip you to accurately read, assess, and answer test questions.

A large part of the guide is devoted to showing you what content to expect on the exam and to helping you better understand that content. In this guide are practice test questions so that you can see how well you have grasped the content. Then, answer explanations are provided so that you can understand why you missed certain questions.

Don't try to cram the night before you take your exam. This is not a wise strategy for a few reasons. First, your retention of the information will be low. Your time would be better used by reviewing information you already know rather than trying to learn a lot of new information. Second, you will likely become stressed as you try to gain a large amount of knowledge in a short amount of time. Third, you will be depriving yourself of sleep. So be sure to go to bed at a reasonable time the night before. Being well-rested helps you focus and remain calm.

Be sure to eat a substantial breakfast the morning of the exam. If you are taking the exam in the afternoon, be sure to have a good lunch as well. Being hungry is distracting and can make it difficult to focus. You have hopefully spent lots of time preparing for the exam. Don't let an empty stomach get in the way of success!

When travelling to the testing center, leave earlier than needed. That way, you have a buffer in case you experience any delays. This will help you remain calm and will keep you from missing your appointment time at the testing center.

Be sure to pace yourself during the exam. Don't try to rush through the exam. There is no need to risk performing poorly on the exam just so you can leave the testing center early. Allow yourself to use all of the allotted time if needed.

Remain positive while taking the exam even if you feel like you are performing poorly. Thinking about the content you should have mastered will not help you perform better on the exam.

Once the exam is complete, take some time to relax. Even if you feel that you need to take the exam again, you will be well served by some down time before you begin studying again. It's often easier to convince yourself to study if you know that it will come with a reward!

Test-Taking Strategies

1. Predicting the Answer

When you feel confident in your preparation for a multiple-choice test, try predicting the answer before reading the answer choices. This is especially useful on questions that test objective factual knowledge. By predicting the answer before reading the available choices, you eliminate the possibility that you will be distracted or led astray by an incorrect answer choice. You will feel more confident in your selection if you read the question, predict the answer, and then find your prediction among the answer choices. After using this strategy, be sure to still read all of the answer choices carefully and completely. If you feel unprepared, you should not attempt to predict the answers. This would be a waste of time and an opportunity for your mind to wander in the wrong direction.

2. Reading the Whole Question

Too often, test takers scan a multiple-choice question, recognize a few familiar words, and immediately jump to the answer choices. Test authors are aware of this common impatience, and they will sometimes prey upon it. For instance, a test author might subtly turn the question into a negative, or he or she might redirect the focus of the question right at the end. The only way to avoid falling into these traps is to read the entirety of the question carefully before reading the answer choices.

3. Looking for Wrong Answers

Long and complicated multiple-choice questions can be intimidating. One way to simplify a difficult multiple-choice question is to eliminate all of the answer choices that are clearly wrong. In most sets of answers, there will be at least one selection that can be dismissed right away. If the test is administered on paper, the test taker could draw a line through it to indicate that it may be ignored; otherwise, the test taker will have to perform this operation mentally or on scratch paper. In either case, once the obviously incorrect answers have been eliminated, the remaining choices may be considered. Sometimes identifying the clearly wrong answers will give the test taker some information about the correct answer. For instance, if one of the remaining answer choices is a direct opposite of one of the eliminated answer choices, it may well be the correct answer. The opposite of obviously wrong is obviously right! Of course, this is not always the case. Some answers are obviously incorrect simply because they are irrelevant to the question being asked. Still, identifying and eliminating some incorrect answer choices is a good way to simplify a multiple-choice question.

4. Don't Overanalyze

Anxious test takers often overanalyze questions. When you are nervous, your brain will often run wild, causing you to make associations and discover clues that don't actually exist. If you feel that this may be a problem for you, do whatever you can to slow down during the test. Try taking a deep breath or counting to ten. As you read and consider the question, restrict yourself to the particular words used by the author. Avoid thought tangents about what the author *really* meant, or what he or she was *trying* to say. The only things that matter on a multiple-choice test are the words that are actually in the question. You must avoid reading too much into a multiple-choice question, or supposing that the writer meant something other than what he or she wrote.

5. No Need for Panic

It is wise to learn as many strategies as possible before taking a multiple-choice test, but it is likely that you will come across a few questions for which you simply don't know the answer. In this situation, avoid panicking. Because most multiple-choice tests include dozens of questions, the relative value of a single wrong answer is small. As much as possible, you should compartmentalize each question on a multiple-choice test. In other words, you should not allow your feelings about one question to affect your success on the others. When you find a question that you either don't understand or don't know how to answer, just take a deep breath and do your best. Read the entire question slowly and carefully. Try rephrasing the question a couple of different ways. Then, read all of the answer choices carefully. After eliminating obviously wrong answers, make a selection and move on to the next question.

6. Confusing Answer Choices

When working on a difficult multiple-choice question, there may be a tendency to focus on the answer choices that are the easiest to understand. Many people, whether consciously or not, gravitate to the answer choices that require the least concentration, knowledge, and memory. This is a mistake. When you come across an answer choice that is confusing, you should give it extra attention. A question might be confusing because you do not know the subject matter to which it refers. If this is the case, don't eliminate the answer before you have affirmatively settled on another. When you come across an answer choice of this type, set it aside as you look at the remaining choices. If you can confidently assert that one of the other choices is correct, you can leave the confusing answer aside. Otherwise, you will need to take a moment to try to better understand the confusing answer choice. Rephrasing is one way to tease out the sense of a confusing answer choice.

7. Your First Instinct

Many people struggle with multiple-choice tests because they overthink the questions. If you have studied sufficiently for the test, you should be prepared to trust your first instinct once you have carefully and completely read the question and all of the answer choices. There is a great deal of research suggesting that the mind can come to the correct conclusion very quickly once it has obtained all of the relevant information. At times, it may seem to you as if your intuition is working faster even than your reasoning mind. This may in fact be true. The knowledge you obtain while studying may be retrieved from your subconscious before you have a chance to work out the associations that support it. Verify your instinct by working out the reasons that it should be trusted.

8. Key Words

Many test takers struggle with multiple-choice questions because they have poor reading comprehension skills. Quickly reading and understanding a multiple-choice question requires a mixture of skill and experience. To help with this, try jotting down a few key words and phrases on a piece of scrap paper. Doing this concentrates the process of reading and forces the mind to weigh the relative importance of the question's parts. In selecting words and phrases to write down, the test taker thinks about the question more deeply and carefully. This is especially true for multiple-choice questions that are preceded by a long prompt.

9. Subtle Negatives

One of the oldest tricks in the multiple-choice test writer's book is to subtly reverse the meaning of a question with a word like *not* or *except*. If you are not paying attention to each word in the question, you can easily be led astray by this trick. For instance, a common question format is, "Which of the following is...?" Obviously, if the question instead is, "Which of the following is not...?," then the answer will be quite different. Even worse, the test makers are aware of the potential for this mistake and will include one answer choice that would be correct if the question were not negated or reversed. A test taker who misses the reversal will find what he or she believes to be a correct answer and will be so confident that he or she will fail to reread the question and discover the original error. The only way to avoid this is to practice a wide variety of multiple-choice questions and to pay close attention to each and every word.

10. Reading Every Answer Choice

It may seem obvious, but you should always read every one of the answer choices! Too many test takers fall into the habit of scanning the question and assuming that they understand the question because they recognize a few key words. From there, they pick the first answer choice that answers the question they believe they have read. Test takers who read all of the answer choices might discover that one of the latter answer choices is actually *more* correct. Moreover, reading all of the answer choices can remind you of facts related to the question that can help you arrive at the correct answer. Sometimes, a misstatement or incorrect detail in one of the latter answer choices will trigger your memory of the subject and will enable you to find the right answer. Failing to read all of the answer choices is like not reading all of the items on a restaurant menu: you might miss out on the perfect choice.

11. Spot the Hedges

One of the keys to success on multiple-choice tests is paying close attention to every word. This is never truer than with words like almost, most, some, and sometimes. These words are called "hedges" because they indicate that a statement is not totally true or not true in every place and time. An absolute statement will contain no hedges, but in many subjects, the answers are not always straightforward or absolute. There are always exceptions to the rules in these subjects. For this reason, you should favor those multiple-choice questions that contain hedging language. The presence of qualifying words indicates that the author is taking special care with his or her words, which is certainly important when composing the right answer. After all, there are many ways to be wrong, but there is only one way to be right! For this reason, it is wise to avoid answers that are absolute when taking a multiple-choice test. An absolute answer is one that says things are either all one way or all another. They often include words like *every*, *always*, *best*, and *never*. If you are taking a multiple-choice test in a subject that doesn't lend itself to absolute answers, be on your guard if you see any of these words.

12. Long Answers

In many subject areas, the answers are not simple. As already mentioned, the right answer often requires hedges. Another common feature of the answers to a complex or subjective question are qualifying clauses, which are groups of words that subtly modify the meaning of the sentence. If the question or answer choice describes a rule to which there are exceptions or the subject matter is complicated, ambiguous, or confusing, the correct answer will require many words in order to be expressed clearly and accurately. In essence, you should not be deterred by answer choices that seem excessively long. Oftentimes, the author of the text will not be able to write the correct answer without offering some qualifications and modifications. Your job is to read the answer choices thoroughly and completely and to select the one that most accurately and precisely answers the question.

13. Restating to Understand

Sometimes, a question on a multiple-choice test is difficult not because of what it asks but because of how it is written. If this is the case, restate the question or answer choice in different words. This process serves a couple of important purposes. First, it forces you to concentrate on the core of the question. In order to rephrase the question accurately, you have to understand it well. Rephrasing the question will concentrate your mind on the key words and ideas. Second, it will present the information to your mind in a fresh way. This process may trigger your memory and render some useful scrap of information picked up while studying.

14. True Statements

Sometimes an answer choice will be true in itself, but it does not answer the question. This is one of the main reasons why it is essential to read the question carefully and completely before proceeding to the answer choices. Too often, test takers skip ahead to the answer choices and look for true statements. Having found one of these, they are content to select it without reference to the question above. Obviously, this provides an easy way for test makers to play tricks. The savvy test taker will always read the entire question before turning to the answer choices. Then, having settled on a correct answer choice, he or she will refer to the original question and ensure that the selected answer is relevant. The mistake of choosing a correct-but-irrelevant answer choice is especially common on questions related to specific pieces of objective knowledge. A prepared test taker will have a wealth of factual knowledge at his or her disposal, and should not be careless in its application.

15. No Patterns

One of the more dangerous ideas that circulates about multiple-choice tests is that the correct answers tend to fall into patterns. These erroneous ideas range from a belief that B and C are the most common right answers, to the idea that an unprepared test-taker should answer "A-B-A-C-A-D-A-B-A." It cannot be emphasized enough that pattern-seeking of this type is exactly the WRONG way to approach a multiple-choice test. To begin with, it is highly unlikely that the test maker will plot the correct answers according to some predetermined pattern. The questions are scrambled and delivered in a random order. Furthermore, even if the test maker was following a pattern in the assignation of correct answers, there is no reason why the test taker would know which pattern he or she was using. Any attempt to discern a pattern in the answer choices is a waste of time and a distraction from the real work of taking the test. A test taker would be much better served by extra preparation before the test than by reliance on a pattern in the answers.

FREE DVD OFFER

Don't forget that doing well on your exam includes both understanding the test content and understanding how to use what you know to do well on the test. We offer a completely FREE Test Taking Tips DVD that covers world class test taking tips that you can use to be even more successful when you are taking your test.

All that we ask is that you email us your feedback about your study guide. To get your **FREE Test Taking Tips DVD**, email freedvd@studyguideteam.com with "FREE DVD" in the subject line and the following information in the body of the email:

- The title of your study guide.
- Your product rating on a scale of 1-5, with 5 being the highest rating.
- Your feedback about the study guide. What did you think of it?
- Your full name and shipping address to send your free DVD.

Introduction

Function of the Test

The purpose of the Common Core Math Grade 8 test to assess eighth grade students' mastery of the mathematics standards that have been identified as those appropriate for the grade level. The material tested on the Common Core Math Grade 8 should be addressed by in-class mathematics education and the curriculum designed by the school the student attends. Achieving a high score on the test demonstrates that a given student understands the mathematics concepts deemed important for grade-level academic success and that instructional delivery of the skills and concepts was effective.

Test Administration

Over 40 states have adopted the Common Core Standards. Each one of these states develops and uses their own version of a Common Core Math test and as such, administration varies from state to state. The test is usually split into two or three testing sessions that take place during school hours in place of normal instructional. Students who finish early are usually instructed to read quietly or do other independent work.

Test Format

The test format for Common Core Math tests varies from state to state; however, the material tested is consistent across states. The number and type of questions varies. The 2018 New York State Grade 8 Math test contained 48 questions: 40 multiple-choice questions, and 8 constructed-response questions. Testing was divided into two sessions. In contrast, Louisiana's test was administered in three sessions. It contained 60 multiple-choice questions (30 without the use of a calculator and 30 where a calculator was permitted) and 4 extended-response question.

The Grade 8 math exam focuses on five primary areas: The Number System, Expressions and Equations, Functions, Geometry, and Statistics and Probability. Within these topics, there is a focus on rational and irrational numbers, exponents and roots, proportional relationships, linear equations and functions, congruence and similarity, the Pythagorean theorem, volume, and trends in data.

The following table shows the domains of mathematics assessed on the exam the number of standards tested for each:

Domain	Number of Standards Tested
The Number System	2
Expressions and Equations	8
Functions	5
Geometry	9
Statistics and Probability	4

Scoring

Because each state has the latitude to design and administer their own version of the test, scores cannot be easily compared between states.

The Number System

Rational and Irrational Numbers

Rational and Irrational Numbers

All real numbers can be separated into two groups: rational and irrational numbers. **Rational numbers** are any numbers that can be written as a fraction, such as $\frac{1}{3}, \frac{7}{4}$, and -25. Alternatively, **irrational numbers** are those that cannot be written as a fraction, such as numbers with never-ending, non-repeating decimal values. Many irrational numbers result from taking roots, such as $\sqrt{2}$ or $\sqrt{3}$. An irrational number may be written as:

34.5684952...

The ellipsis (...) represents the line of numbers after the decimal that does not repeat and is never-ending.

When rational and irrational numbers interact, there are different types of number outcomes. For example, when adding or multiplying two rational numbers, the result is a rational number. No matter what two fractions are added or multiplied together, the result can always be written as a fraction. The following expression shows two rational numbers multiplied together:

$$\frac{3}{8} \times \frac{4}{7} = \frac{12}{56}$$

The product of these two fractions is another fraction that can be simplified to $\frac{3}{14}$.

As another interaction, rational numbers added to irrational numbers will always result in irrational numbers. No part of any fraction can be added to a never-ending, non-repeating decimal to make a rational number. The same result is true when multiplying a rational and irrational number. Taking a fractional part of a never-ending, non-repeating decimal will always result in another never-ending, non-repeating decimal. An example of the product of rational and irrational numbers is shown in the following expression: $2\times \sqrt{7}$.

The last type of interaction concerns two irrational numbers, where the sum or product may be rational or irrational depending on the numbers being used. The following expression shows a rational sum from two irrational numbers:

$$\sqrt{3} + (6 - \sqrt{3}) = 6$$

The product of two irrational numbers can be rational or irrational. A rational result can be seen in the following expression:

$$\sqrt{2} \times \sqrt{8} = \sqrt{2 \times 8} = \sqrt{16} = 4$$

An irrational result can be seen in the following:

$$\sqrt{3} \times \sqrt{2} = \sqrt{6}$$

Approximating the Value of an Irrational Number and Locating its Approximation on a Number Line

All real numbers can be separated into two groups: rational and irrational numbers. **Rational numbers** are any numbers that can be written as a fraction, such as $\frac{1}{3}, \frac{7}{4}$, and -25. Alternatively, **irrational numbers** are those that cannot be written as a fraction, such as numbers with never-ending, non-repeating decimal values. Many irrational numbers result from taking roots, such as $\sqrt{2}$ or $\sqrt{3}$. An irrational number may be written as:

34.5684952...

The **ellipsis** (...) represents the line of numbers after the decimal that does not repeat and is never-ending.

As mentioned, an irrational number is any number that cannot be written as a fraction or a non-repeating or terminal decimal. Examples of irrational numbers involve roots that do not work out nicely, such as $\sqrt{15}$. Sometimes it is helpful to approximate irrational numbers because, for example, you might not be able to calculate in your head what the decimal form of $\sqrt{15}$ is. The way to approximate such an irrational number is to locate the closest perfect square on either side of the radicand 15. A perfect square is a positive integer that when you take the square root of that value, you obtain a positive integer. For example, $\sqrt{15}$ lies between $\sqrt{9}$ = 3 and $\sqrt{16}$ = 4. It is closest to $\sqrt{16} = 4$, so we estimate that $\sqrt{15}$ is close to 4, but less than 4. It is actually equal to 3.87, rounded to two decimal places.

When rational and irrational numbers interact, there are different types of number outcomes. For example, when adding or multiplying two rational numbers, the result is a rational number. No matter what two fractions are added or multiplied together, the result can always be written as a fraction. The following expression shows two rational numbers multiplied together:

$$\frac{3}{8} \times \frac{4}{7} = \frac{12}{56}$$

The product of these two fractions is another fraction that can be simplified to $\frac{3}{14}$.

As another interaction, rational numbers added to irrational numbers will always result in irrational numbers. No part of any fraction can be added to a never-ending, non-repeating decimal to make a rational number. The same result is true when multiplying a rational and irrational number. Taking a fractional part of a never-ending, non-repeating decimal will always result in another never-ending, non-repeating decimal. An example of the product of rational and irrational numbers is shown in the following expression: $2\times \sqrt{7}$.

The last type of interaction concerns two irrational numbers, where the sum or product may be rational or irrational depending on the numbers used. The following expression shows a rational sum from two irrational numbers:

$$\sqrt{3} + \left(6 - \sqrt{3}\right) = 6$$

The product of two irrational numbers can be rational or irrational. A rational result can be seen in the following expression:

$$\sqrt{2} \times \sqrt{8} = \sqrt{2 \times 8} = \sqrt{16} = 4$$

An irrational result can be seen in the following:

$$\sqrt{3} \times \sqrt{2} = \sqrt{6}$$

A **number line** typically consists of integers (...3,2,1,0,-1,-2,-3...), and is used to visually represent the value of a number. Although number lines are generally used with rational numbers, they can be used to represent the approximate value of an irrational number as well. Each rational number has a distinct position on the line determined by comparing its value with the displayed values on the line. For example, if plotting -1.5 on the number line below, it is necessary to recognize that the value of -1.5 is .5 less than -1 and .5 greater than -2. Therefore, -1.5 is plotted halfway between -1 and -2. The approximate value (rounded) of an imaginary number can be plotted much the same way. It should be noted that the more precise the approximation, the closer the plotted value will be to the true value. Most number lines are subdivided into integers so that the interval between two labeled marks is equal to 1. Therefore, some degree of approximation is often used, even when plotting rational numbers. For example, on the number line below, the distance between each hash mark is 1.

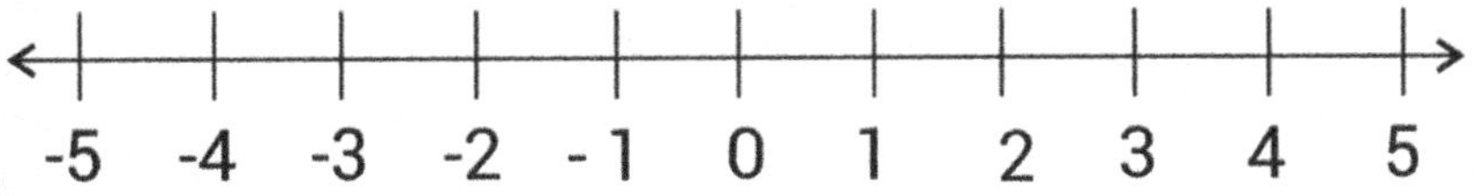

Ordering Rational Numbers

A common question type asks to order rational numbers from least to greatest or greatest to least. The numbers will come in a variety of formats, including decimals, percentages, roots, fractions, and whole numbers. These questions test for knowledge of different types of numbers and the ability to determine their respective values.

Before discussing ordering all numbers, let's start with decimals.

To compare decimals and order them by their value, utilize a method similar to that of ordering large numbers.

The main difference is where the comparison will start. Assuming that any numbers to left of the decimal point are equal, the next numbers to be compared are those immediately to the right of the decimal point. If those are equal, then move on to compare the values in the next decimal place to the right.

For example: Which number is greater, 12.35 or 12.38?

Check that the values to the left of the decimal point are equal:

$$12 = 12$$

Next, compare the values of the decimal place to the right of the decimal:

$$12.3 = 12.3$$

Those are also equal in value.

Finally, compare the value of the numbers in the next decimal place to the right on both numbers:

12.3**5** and 12.3**8**

Here the 5 is less than the 8, so the final way to express this inequality is:

$$12.35 < 12.38$$

Comparing decimals is regularly exemplified with money because the "cents" portion of money ends in the hundredths place. When paying for gasoline or meals in restaurants, and even in bank accounts, if enough errors are made when calculating numbers to the hundredths place, they can add up to dollars and larger amounts of money over time.

Now that decimal ordering has been explained, let's expand and consider all real numbers. Whether the question asks to order the numbers from greatest to least or least to greatest, the crux of the question is the same—convert the numbers into a common format. Generally, it's easiest to write the numbers as whole numbers and decimals so they can be placed on a number line. Follow these examples to understand this strategy.

1) Order the following rational numbers from greatest to least:

$$\sqrt{36}, 0.65, 78\%, \frac{3}{4}, 7, 90\%, \frac{5}{2}$$

Of the seven numbers, the whole number (7) and decimal (0.65) are already in an accessible form, so concentrate on the other five.

First, the square root of 36 equals 6. (If the test asks for the root of a non-perfect root, determine which two whole numbers the root lies between.) Next, convert the percentages to decimals. A percentage means "per hundred," so this conversion requires moving the decimal point two places to the left, leaving 0.78 and 0.9. Lastly, evaluate the fractions:

$$\frac{3}{4} = \frac{75}{100} = 0.75\,; \frac{5}{2} = 2\frac{1}{2} = 2.5$$

Now, the only step left is to list the numbers in the request order:

$$7, \sqrt{36}, \frac{5}{2}, 90\%, 78\%, \frac{3}{4}, 0.65$$

2) Order the following rational numbers from least to greatest:

$$2.5, \sqrt{9}, \text{-}10.5, 0.853, 175\%, \sqrt{4}, \frac{4}{5}$$

$$\sqrt{9} = 3 \qquad 175\% = 1.75$$

$$\sqrt{4} = 2 \qquad \frac{4}{5} = 0.8$$

From least to greatest, the answer is:

$$\text{-}10.5, \frac{4}{5}, 0.853, 175\%, \sqrt{4}, 2.5, \sqrt{9}$$

Expressions and Equations

Working with Radicals and Exponents

Properties of Exponents

An **exponent** is an operation used as shorthand for a number multiplied or divided by itself for a defined number of times.

$$3^7 = 3 \times 3 \times 3 \times 3 \times 3 \times 3 \times 3$$

In this example, the 3 is called the **base**, and the 7 is called the **exponent**. The exponent is typically expressed as a superscript number near the upper right side of the base but can also be identified as the number following a caret symbol (^). This operation is verbally expressed as "3 to the 7th power" or "3 raised to the power of 7." Common exponents are 2 and 3. A base raised to the power of 2 is referred to as having been "squared," while a base raised to the power of 3 is referred to as having been "cubed."

Several special rules apply to exponents. First, the **Zero Power Rule** finds that any number raised to the zero power equals 1. For example, 100^0, 2^0, $(-3)^0$and 0^0 all equal 1 because the bases are raised to the zero power.

Second, exponents can be negative. With negative exponents, the equation is expressed as a fraction, as in the following example:

$$3^{-7} = \frac{1}{3^7} = \frac{1}{3 \text{ x } 3 \text{ x } 3 \text{ x } 3 \text{ x } 3 \text{ x } 3 \text{ x } 3}$$

Third, the **Power Rule** concerns exponents being raised by another exponent. When this occurs, the exponents are multiplied by each other:

$$(x^2)^3 = x^6 = (x^3)^2$$

Fourth, when multiplying two exponents with the same base, the **Product Rule** requires that the base remains the same and the exponents are added. For example, $a^x \text{x } a^y = a^{x+y}$. Since addition and multiplication are commutative, the two terms being multiplied can be in any order.

$$x^3x^5 = x^{3+5} = x^8 = x^{5+3} = x^5x^3$$

Fifth, when dividing two exponents with the same base, the **Quotient Rule** requires that the base remains the same, but the exponents are subtracted. So, $a^x \div a^y = a^{x-y}$. Since subtraction and division are not commutative, the two terms must remain in order.

$$x^5x^{-3} = x^{5-3} = x^2 = x^5 \div x^3 = \frac{x^5}{x^3}$$

Additionally, 1 raised to any power is still equal to 1, and any number raised to the power of 1 is equal to itself. In other words, $a^1 = a$ and $14^1 = 14$.

Exponents play an important role in scientific notation to present extremely large or small numbers as follows: $a \times 10^b$. To write the number in scientific notation, the decimal is moved until there is only one digit on the left side of the decimal point, indicating that the number *a* has a value between 1 and

10. The number of times the decimal moves indicates the exponent to which 10 is raised, here represented by b. If the decimal moves to the left, then b is positive, but if the decimal moves to the right, then b is negative. The following examples demonstrate these concepts:

$$3{,}050 = 3.05 \times 10^3$$

$$-777 = -7.77 \times 10^2$$

$$0.000123 = 1.23 \times 10^{-4}$$

$$-0.0525 = -5.25 \times 10^{-2}$$

As a summary, there are a few rules for working with exponents. For any numbers a, b, m, n, the following hold true:

$$a^1 = a$$

$$1^a = 1$$

$$a^0, = 1$$

$$a^m \times a^n = a^{m+n}$$

$$a^m \div a^n = a^{m-n}$$

$$(a^m)^n = a^{m \times n}$$

$$(a \times b)^m = a^m \times b^m$$

$$(a \div b)^m = a^m \div b^m$$

Any number, including a fraction, can be an exponent. The same rules apply.

Roots

The **square root** symbol is expressed as $\sqrt{\ }$ and is commonly known as the radical. Taking the root of a number is the inverse operation of multiplying that number by itself some number of times. For example, squaring the number 7 is equal to 7×7, or 49. Finding the square root is the opposite of finding an exponent, as the operation seeks a number that when multiplied by itself, equals the number in the square root symbol.

For example, $\sqrt{36}$ = 6 because 6 multiplied by 6 equals 36. Note, the square root of 36 is also -6 since:

$$-6 \times -6 = 36$$

This can be indicated using a **plus/minus** symbol like this: ±6. However, square roots are often just expressed as a positive number for simplicity, with it being understood that the true value can be either positive or negative.

Perfect squares are numbers with whole number square roots. The list of perfect squares begins with:

0, 1, 4, 9, 16, 25, 36, 49, 64, 81, and 100

Determining the square root of imperfect squares requires a calculator to reach an exact figure. It's possible, however, to approximate the answer by finding the two perfect squares that the number fits between. For example, the square root of 40 is between 6 and 7 since the squares of those numbers are 36 and 49, respectively.

Square roots are the most common root operation. If the radical doesn't have a number to the upper left of the symbol $\sqrt{}$, then it's a **square root**. Sometimes a radical includes a number in the upper left, like $\sqrt[3]{27}$, as in the other common root type—the **cube root**. Complicated roots, like the cube root, often require a calculator.

Using Powers of 10 to Estimate Quantities

A single digit integer can be multiplied by power of 10 to estimate a very large or very small quantity. A power of 10 refers to 10 raised to a given exponent such as 10^0, 10^1, 10^2, 10^3, etc. For the number 10^3, 10 is the base and 3 is the exponent. A base raised by an exponent represents how many times the base is multiplied by itself. Therefore, $10^1 = 10$, $10^2 = 10 \times 10 = 100$, $10^3 = 10 \times 10 \times 10 = 1{,}000$, $10^4 = 10 \times 10 \times 10 \times 10 = 10{,}000$, etc. Any base with a zero exponent equals one.

The power of 10 by which the a single-digit number is multiplied corresponds to the number of zeroes following the digit when expressing its value in standard form. For example, 7×10^4 is equivalent to 70,000 or 7 followed by four zeros. These numbers can be used for estimations such as the population of the world being approximately 7×10^9, or 7 billion, while the population of New York City can be approximated at 8×10^6 for 8 million.

Recall that in general, **estimation** is finding a value that is close to a solution but is not the exact answer. For example, if there are values in the thousands to be multiplied, then each value can be estimated to the nearest thousand and the calculation performed. This value provides an approximate solution that can be determined very quickly.

Rounding is the process of either bumping a number up or down, based on a specified place value. First, the place value is specified. Then, the digit to its right is looked at. For example, if rounding to the nearest hundreds place, the digit in the tens place is used. If it is a 0, 1, 2, 3, or 4, the digit being rounded to is left alone. If it is a 5, 6, 7, 8 or 9, the digit being rounded to is increased by one. All other digits before the decimal point are then changed to zeros, and the digits in decimal places are dropped. If a decimal place is being rounded to, all subsequent digits are just dropped. For example, if 845,231.45 was to be rounded to the nearest thousands place, the answer would be 845,000. The 5 would remain the same due to the 2 in the hundreds place. Also, if 4.567 was to be rounded to the nearest tenths place, the answer would be 4.6. The 5 increased to 6 due to the 6 in the hundredths place, and the rest of the decimal is dropped.

Sometimes when performing operations such as multiplying numbers, the result can be estimated by rounding. For example, to estimate the value of 11.2×2.01, each number can be rounded to the nearest integer. This will yield a result of 22.

Rounding numbers helps with estimation because it changes the given number to a simpler, although less accurate, number than the exact given number. Rounding allows for easier calculations, which estimate the results of using the exact given number. The accuracy of the estimate and ease of use depends on the place value to which the number is rounded. Rounding numbers consists of:

- determining what place value the number is being rounded to
- examining the digit to the right of the desired place value to decide whether to round up or keep the digit, and
- replacing all digits to the right of the desired place value with zeros.

To round 746,311 to the nearest ten thousand, the digit in the ten thousands place should be located first. In this case, this digit is 4 (7<u>4</u>6,311). Then, the digit to its right is examined. If this digit is 5 or greater, the number will be rounded up by increasing the digit in the desired place by one. If the digit to the right of the place value being rounded is 4 or less, the number will be kept the same. For the given example, the digit being examined is a 6, which means that the number will be rounded up by increasing the digit to the left by one. Therefore, the digit 4 is changed to a 5. Finally, to write the rounded number, any digits to the left of the place value being rounded remain the same and any to its right are replaced with zeros. For the given example, rounding 746,311 to the nearest ten thousand will produce 750,000. To round 746,311 to the nearest hundred, the digit to the right of the three in the hundreds place is examined to determine whether to round up or keep the same number. In this case, that digit is a 1, so the number will be kept the same and any digits to its right will be replaced with zeros. The resulting rounded number is 746,300.

Rounding place values to the right of the decimal follows the same procedure, but digits being replaced by zeros can simply be dropped. To round 3.752891 to the nearest thousandth, the desired place value is located (3.75<u>2</u>891) and the digit to the right is examined. In this case, the digit 8 indicates that the number will be rounded up, and the 2 in the thousandths place will increase to a 3. Rounding up and replacing the digits to the right of the thousandths place produces 3.753000 which is equivalent to 3.753. Therefore, the zeros are not necessary and the rounded number should be written as 3.753.

When rounding up, if the digit to be increased is a 9, the digit to its left is increased by 1 and the digit in the desired place value is changed to a zero. For example, the number 1,598 rounded to the nearest ten is 1,600. Another example shows the number 43.72961 rounded to the nearest thousandth is 43.730 or 43.73.

Mental math should always be considered as problems are worked through, and the ability to work through problems in one's head helps save time. If a problem is simple enough, such as $15 + 3 = 18$, it should be completed mentally. The ability to do this will increase once addition and subtraction in higher place values are grasped. Also, mental math is important in multiplication and division. The times tables multiplying all numbers from 1 to 12 should be memorized. This will allow for division within those numbers to be memorized as well. For example, we should know easily that $121 \div 11 = 11$ because it should be memorized that $11 \times 11 = 121$.

Here is the multiplication table to be memorized:

x	1	2	3	4	5	6	7	8	9	10	11	12	13	14	15
1	1	2	3	4	5	6	7	8	9	10	11	12	13	14	15
2	2	4	6	8	10	12	14	16	18	20	22	24	26	28	30
3	3	6	9	12	15	18	21	24	27	30	33	36	39	42	45
4	4	8	12	16	20	24	28	32	36	40	44	48	52	56	60
5	5	10	15	20	25	30	35	40	45	50	55	60	65	70	75
6	6	12	18	24	30	36	42	48	54	60	66	72	78	84	90
7	7	14	21	28	35	42	49	56	63	70	77	84	91	98	105
8	8	16	24	32	40	48	56	64	72	80	88	96	104	112	120
9	9	18	27	36	45	54	63	72	81	90	99	108	117	126	135
10	10	20	30	40	50	60	70	80	90	100	110	120	130	140	150
11	11	22	33	44	55	66	77	88	99	110	121	132	143	154	165
12	12	24	36	48	60	72	84	96	108	120	132	144	156	168	180
13	13	26	39	52	65	78	91	104	117	130	143	156	169	182	195
14	14	28	42	56	70	84	98	112	126	140	154	168	182	196	210
15	15	30	45	60	75	90	105	120	135	150	165	180	195	210	225

The values along the diagonal of the table consist of **perfect squares**. A perfect square is a number that represents a product of two equal integers.

Converting Between Standard Decimal Notation and Scientific Notation

The **decimal system** is a way of writing out numbers that uses ten different numerals: 0, 1, 2, 3, 4, 5, 6, 7, 8, and 9. This is also called a "base ten" or "base 10" system. Other bases are also used. For example, computers work with a base of 2. This means they only use the numerals 0 and 1.

The **decimal place** denotes how far to the right of the decimal point a numeral is. The first digit to the right of the decimal point is in the *tenths* place. The next is the **hundredths**. The third is the **thousandths**.

So, 3.142 has a 1 in the tenths place, a 4 in the hundredths place, and a 2 in the thousandths place.

The **decimal point** is a period used to separate the **ones** place from the **tenths** place when writing out a number as a decimal.

A **decimal number** is a number written out with a decimal point instead of as a fraction, for example, 1.25 instead of $\frac{5}{4}$. Depending on the situation, it can sometimes be easier to work with fractions and sometimes easier to work with decimal numbers.

A decimal number is **terminating** if it stops at some point. It is called **repeating** if it never stops but repeats a pattern over and over. It is important to note that every rational number can be written as a terminating decimal or as a repeating decimal.

Scientific Notation is used to represent numbers that are either very small or very large. For example, the distance to the Sun is approximately 150,000,000,000 meters. Instead of writing this number with so many zeros, it can be written in scientific notation as 1.5×10^{11}meters. The same is true for very small numbers, but the exponent becomes negative. If the mass of a human cell is 0.000000000001 kilograms, that measurement can be easily represented by 1.0×10^{-12} kilograms. In both situations, scientific notation makes the measurement easier to read and understand. Each number is translated to an expression with one digit in the tens place multiplied by an expression corresponding to the zeros.

Exponents play an important role in scientific notation to present extremely large or small numbers as follows: $a \times 10^b$. To write the number in scientific notation, the decimal is moved until there is only one digit on the left side of the decimal point, indicating that the number *a* has a value between 1 and 10. The number of times the decimal moves indicates the exponent to which 10 is raised, here represented by *b*. If the decimal moves to the left, then *b* is positive, but if the decimal moves to the right, then *b* is negative.

See the following examples:

$$3{,}050 = 3.05 \times 10^3$$

$$-777 = -7.77 \times 10^2$$

$$0.000123 = 1.23 \times 10^{-4}$$

$$-0.0525 = -5.25 \times 10^{-2}$$

When two measurements are given and both involve scientific notation, it is important to know how these interact with each other:

- In addition and subtraction, the exponent on the ten must be the same before any operations are performed on the numbers. For example, $(1.3 \times 10^4) + (3.0 \times 10^3)$ cannot be added until one of the exponents on the ten is changed. The 3.0×10^3 can be changed to 0.3×10^4, then the 1.3 and 0.3 can be added. The answer comes out to be 1.6×10^4.
- For multiplication, the first numbers can be multiplied and then the exponents on the tens can be added. Once an answer is formed, it may have to be converted into scientific notation again depending on the change that occurred.
- The following is an example of multiplication with scientific notation:

$$(4.5 \times 10^3) \times (3.0 \times 10^{-5}) = 13.5 \times 10^{-2}$$

- Since this answer is not in scientific notation, the decimal is moved over to the left one unit, and 1 is added to the ten's exponent. This results in the final answer: 1.35×10^{-1}.
- For division, the first numbers are divided, and the exponents on the tens are subtracted. Again, the answer may need to be converted into scientific notation form, depending on the type of changes that occurred during the problem.

- **Order of magnitude** relates to scientific notation and is the total count of powers of 10 in a number. For example, there are 6 orders of magnitude in 1,000,000. If a number is raised by an order of magnitude, it is multiplied by 10. Order of magnitude can be helpful in estimating results using very large or small numbers. An answer should make sense in terms of its order of magnitude.

- For example, if area is calculated using two dimensions with 6 orders of magnitude, because area involves multiplication, the answer should have around 12 orders of magnitude. Also, answers can be estimated by rounding to the largest place value in each number. For example, 5,493,302 × 2,523,100 can be estimated by 5 × 3 = 15 with 6 orders of magnitude.

Proportional Relationships, Lines, and Linear Equations

Graphing Proportional Relationships

Rate of change for any line calculates the steepness of the line over a given interval. Rate of change is also known as the **slope** or rise/run. The slope of a linear function is given by the change in *y* divided by the change in *x*. So, the formula looks like this:

$$slope = \frac{y_2 - y_1}{x_2 - x_1}$$

In the graph below, two points are plotted. The first has the coordinates of (0,1), and the second point is (2,3). Remember that the x coordinate is always placed first in coordinate pairs. Work from left to right when identifying coordinates. Thus, the point on the left is point 1 (0,1), and the point on the right is point 2 (2,3).

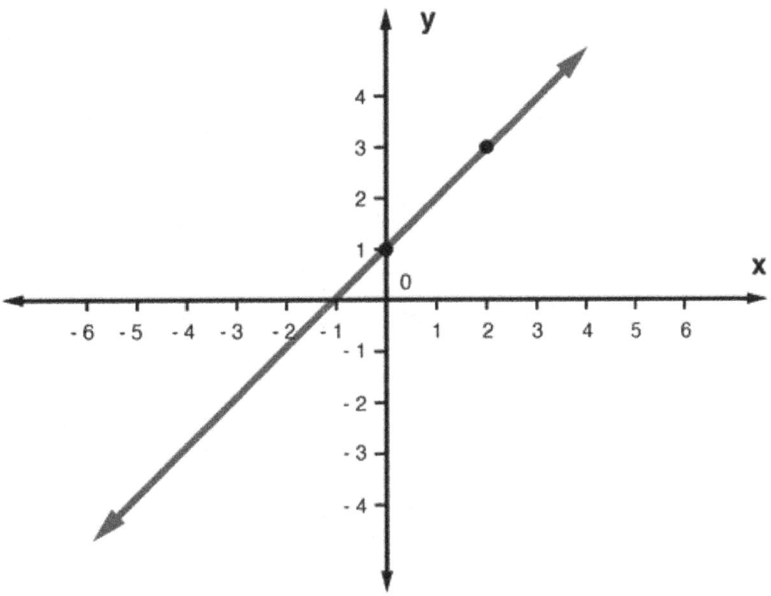

Now we need to just plug those numbers into the equation:

$$slope = \frac{3-1}{2-0}$$

$$slope = \frac{2}{2}$$

$$slope = 1$$

This means that for every increase of 1 for x, y also increased by 1. You can see this in the line. When x equaled 0, y equaled 1, and when x was increased to 1, y equaled 2.

Slope can be thought of as determining the rise over run:

$$slope = \frac{rise}{run}$$

The rise being the change vertically on the y axis and the run being the change horizontally on the x axis.

Graphing and Interpreting Proportional Relationships

Two quantities are **proportional** if there exists a constant ratio between them. The constant, k, is known as the constant of proportionality and proportionality can be expressed in equation form as $y = kx$ or $\frac{y}{x} = k$. Basically, if one quantity is proportional to another, one quantity is a constant multiple of the other. For instance, consider the scenario in which you want to double the number of employees in all of the ice cream stands that you own. This would be represented in equation form as $y = 2x$, where x represents current number of employees and y represents future number of employees. Here is a table that shows 4 scenarios. Notice how the y-value is equal to the x-value doubled.

X	Y
1	2
2	4
3	6
4	8

The graph of the equation $y = 2x$ can be seen here:

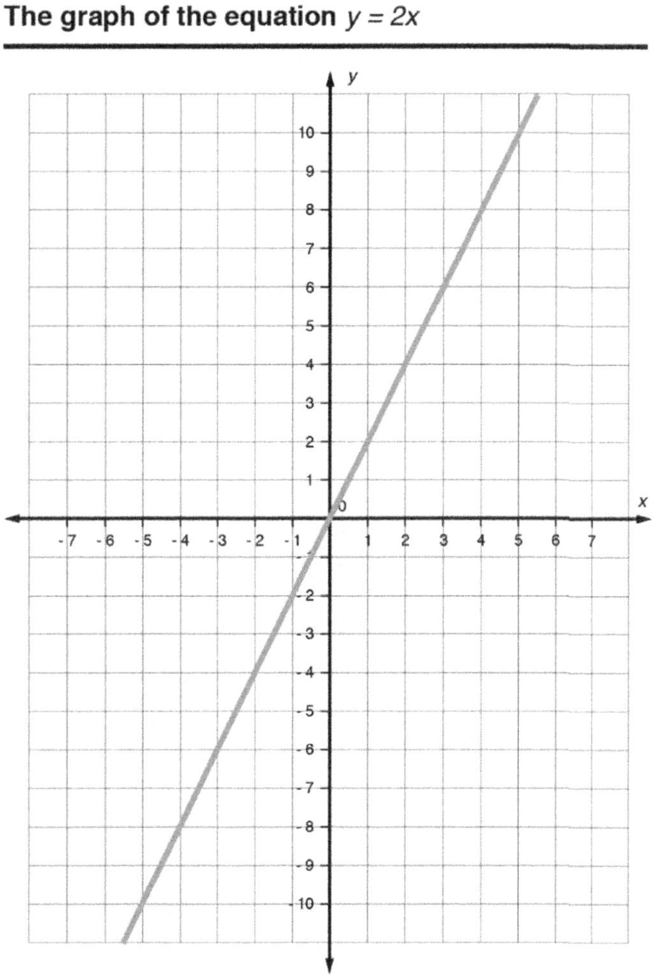

A situation in which a quantity models a linear behavior but is not proportional uses the equation $y = mx + b$. Note that this equation involves another quantity in the equation, b, which does not appear when a quantity is directly proportional. This quantity b is obtained when plugging $x = 0$ into the equation. Consider a situation in which the quantity x is doubled, but also added to the number 2. The equation would be $y = 2x + 2$. A real-world application of this equation could exist within a public transportation setting. Let's say a New York City metro card costs a flat fee of \$2 to purchase the actual card and then costs \$2 per ride. The equation $y = 2x + 2$ models this scenario where x is number of rides that can be taken and y is total cost of purchasing the card.

A table of values could be created from this equation, as seen below:

X	Y
1	4
2	6
3	8
4	10

Notice that each x value is doubled and then added to 2.

The graph of the equation $y = 2x + 2$ is related to the graph of $y = 2x$. The addition of the b value of 2 shifts the graph of $y = 2x$ vertically 2 units. The graph can be seen here:

The graph of the equation $y = 2x + 2$

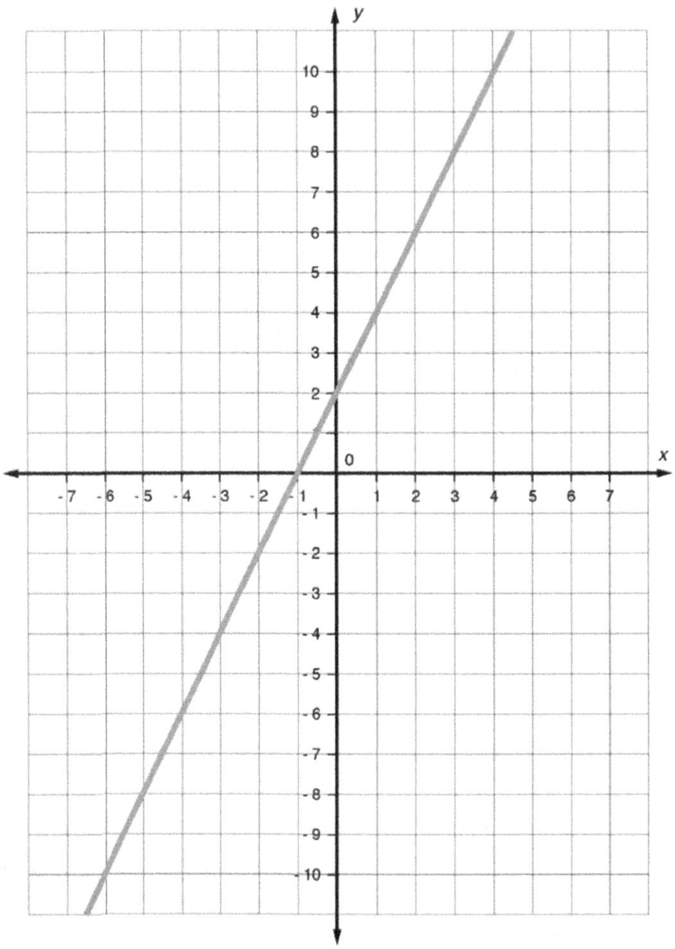

Any linear equation with two variables can be written given a point (ordered pair) and the slope or given two points on a line. An ordered pair gives a set of corresponding values for the two variables (*x* and *y*). As an example, for a scenario involving distance and time, it is given that the person traveled 112.5 miles in 2 ½ hours. Knowing that *x* represents time and *y* represents distance, this information can be written as the ordered pair (2.5, 112.5).

Recall that linear relationships describe the way two quantities change with respect to each other. The relationship is defined as linear because a line is produced if all the sets of corresponding values are graphed on a coordinate grid. When expressing the linear relationship as an equation, the equation is often written in the form $y = mx + b$ (**slope-intercept form**) where *m* and *b* are numerical values and *x* and *y* are variables (for example, $y = 5x + 10$). The slope is the coefficient of x, and the y-intercept is the constant value. The slope of the line containing the same two points is $m = \frac{y_2 - y_1}{x_2 - x_1}$ and is also equal to rise/run. Given a linear equation and the value of either variable (*x* or *y*), the value of the other variable can be determined.

Interpreting Slope as a Constant Rate of Change

The first step to writing a linear model is to identify what the variables represent. A **variable** represents an unknown quantity, and in the case of a linear equation, a specific relationship exists between the two variables (usually *x* and *y*). Within a given scenario, the variables are the two quantities that are changing. The variable *x* is considered the **independent variable** and represents the inputs of a function. The variable *y* is considered the **dependent variable** and represents the outputs of a function. For example, if a scenario describes distance traveled and time traveled, distance would be represented by *y* and time represented by *x*. The distance traveled depends on the time spent traveling (time is independent). If a scenario describes the cost of a cab ride and the distance traveled, the cost would be represented by *y* and the distance represented by *x*. The cost of a cab ride depends on the distance traveled.

The slope represents the rate of change. In linear models, because the slope is constant (a straight line), the rate of change is also constant. In the case of the cab ride, as the distance of the ride increases, so too does the cost in a directly proportional rate.

Rate of change for any line calculates the steepness of the line over a given interval. Rate of change is also known as the **slope** or rise/run. The slope of a linear function is given by the change in *y* divided by the change in *x*. So, the formula looks like this:

$$slope = \frac{y_2 - y_1}{x_2 - x_1}$$

In the graph below, two points are plotted. The first has the coordinates of (0,1) and the second point is (2,3). Remember that the x coordinate is always placed first in coordinate pairs. Work from left to right when identifying coordinates. Thus, the point on the left is point 1 (0,1), and the point on the right is point 2 (2,3).

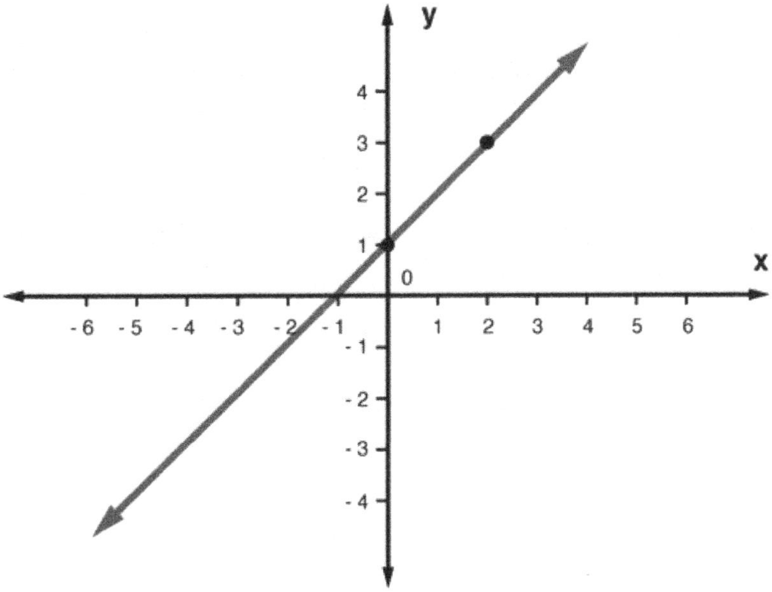

Now we need to just plug those numbers into the equation:

$$slope = \frac{3-1}{2-0}$$

$$slope = \frac{2}{2}$$

$$slope = 1$$

This means that for every increase of 1 for x, y also increased by 1. You can see this in the line. When x equaled 0, y equaled 1, and when x was increased to 1, y equaled 2.

Slope can be thought of as determining the rise over run:

$$slope = \frac{rise}{run}$$

The rise being the change vertically on the y axis and the run being the change horizontally on the x axis.

Graphing Linear Functions

The process for graphing a line depends on the form in which its equation is written: slope-intercept form or standard form.

When an equation is written in **slope-intercept form**, $y = mx + b$, m represents the slope of the line and *b* represents the *y*-intercept. The **y-intercept** is the value of *y* when $x = 0$ and the point at which the graph of the line crosses the *y*-axis. The slope is the rate of change between the variables, expressed as a fraction. The fraction expresses the change in *y* compared to the change in *x*. If the slope is an integer, it should be written as a fraction with a denominator of 1. For example, 5 would be written as 5/1.

To graph a line given an equation in slope-intercept form, the *y*-intercept should first be plotted. For example, to graph $y = -\frac{2}{3}x + 7$, the *y*-intercept of 7 would be plotted on the *y*-axis (vertical axis) at the point (0, 7). Next, the slope would be used to determine a second point for the line. Note that all that is necessary to graph a line is two points on that line. The slope will indicate how to get from one point on the line to another. The slope expresses vertical change (*y*) compared to horizontal change (*x*) and therefore is sometimes referred to as $\frac{rise}{run}$. The numerator indicates the change in the *y* value (move up for positive integers and move down for negative integers), and the denominator indicates the change in the *x* value. For the previous example, using the slope of $-\frac{2}{3}$, from the first point at the *y*-intercept, the second point should be found by counting down 2 and to the right 3. This point would be located at (3, 5).

When an equation is written in **standard form**, $Ax + By = C$, it is easy to identify the *x*- and *y*-intercepts for the graph of the line. Just as the *y*-intercept is the point at which the line intercepts the *y*-axis, the *x*-intercept is the point at which the line intercepts the *x*-axis. At the *y*-intercept, $x = 0$; and at the *x*-intercept, $y = 0$. Given an equation in standard form, $x = 0$ should be used to find the *y*-intercept. Likewise, $y = 0$ should be used to find the *x*-intercept. For example, to graph $3x + 2y = 6$, 0 for *y* results in $3x + 2(0) = 6$. Solving for *y* yields $x = 2$; therefore, an ordered pair for the line is (2, 0). Substituting 0 for *x* results in $3(0) + 2y = 6$. Solving for *y* yields $y = 3$; therefore, an ordered pair for the line is (0, 3).

The two ordered pairs (the *x*- and *y*-intercepts) can be plotted, and a straight line through them can be constructed.

T - chart

x	y
0	3
2	0

Intercepts

x - intercept : (2,0)

y - intercept : (0,3)

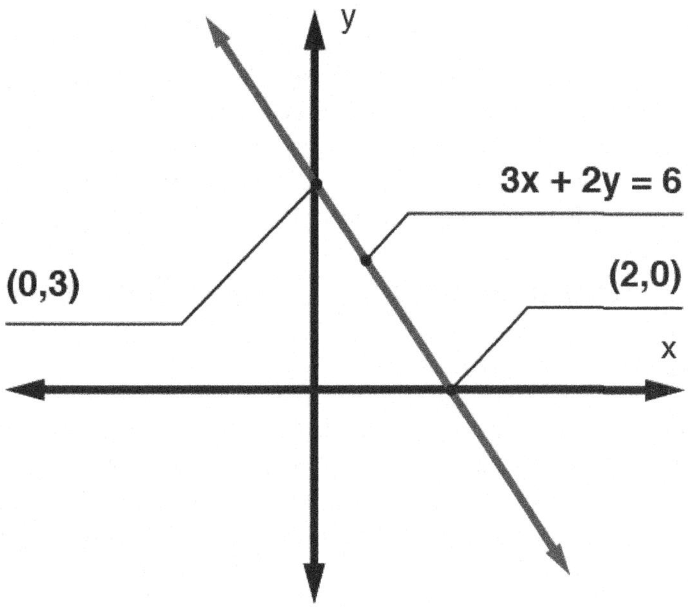

Using Similar Right Triangles to Understand Slope

In order to understand the slope of a line using similar triangles, the triangles must be right triangles. In order to form one of these right triangles, select two points on the line. One side of the triangle connects the two points. Establish a second side of the triangle by drawing a vertical line at the x-value of one of the points, and find the third side of the triangle by drawing a horizontal line at the y-value of the other point. These two new lines will intersect, forming a right triangle. Next, select a third point on the line. In a similar process, draw a right triangle using this new point and one of the other two original points. Then, the two right triangles that have been drawn are similar. An example is shown below:

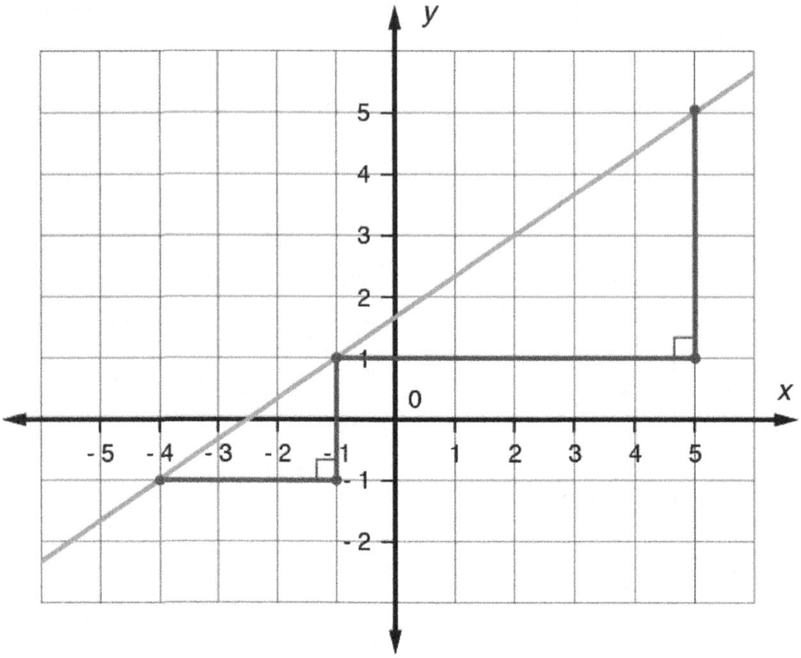

Because the triangles are similar, the ratio of corresponding sides are equal. The **slope** of a line is defined to be the ratio of the change in y over the change in x. Similar triangles show that this ratio will always be the same, regardless of the points chosen. Two similar right triangles can always be built regardless of the three points selected on the line. Therefore, what that means is that any two points on the line can be chosen to determine the slope of the line.

Analyzing and Solving Linear Equations

Solving Linear Equations in One Variable

Linear expressions and equations are concise mathematical statements that can be written to model a variety of scenarios. Questions found pertaining to this topic will contain one variable only. A variable is an unknown quantity, usually denoted by a letter (*x, n, p,* etc.). In the case of linear expressions and equations, the power of the variable (its exponent) is 1. A variable without a visible exponent is raised to the first power.

A linear expression is a statement about an unknown quantity expressed in mathematical symbols. The statement "five times a number added to forty" can be expressed as $5x + 40$. A linear equation is a statement in which two expressions (at least one containing a variable) are equal to each other. The statement "five times a number added to forty is equal to ten" can be expressed as $5x + 40 = 10$. Real-world scenarios can also be expressed mathematically. Consider the following:

> Bob had $20 and Tom had $4. After selling 4 ice cream cones to Bob, Tom has as much money as Bob.

The cost of an ice cream cone is an unknown quantity and can be represented by a variable. The amount of money Bob has after his purchase is four times the cost of an ice cream cone subtracted from his original $20. The amount of money Tom has after his sale is four times the cost of an ice cream cone added to his original $4. This can be expressed as: $20 - 4x = 4x + 4$, where *x* represents the cost of an ice cream cone.

When expressing a verbal or written statement mathematically, it is key to understand words or phrases that can be represented with symbols. The following are examples:

Symbol	**Phrase**
$+$	added to, increased by, sum of, more than
$-$	decreased by, difference between, less than, take away
x	multiplied by, 3 (4, 5 . . .) times as large, product of
$\div$	divided by, quotient of, half (third, etc.) of
$=$	is, the same as, results in, as much as
$x, t, n, etc.$	a number, unknown quantity, value of

Linear equations may take forms other than $y = ax + b$. The most common forms of linear equations are explained below:

1. Standard Form: $Ax + By = C$, in which the slope is given by $m = \frac{-A}{B}$, and the *y*-intercept is given by $\frac{C}{B}$.

2. Slope-Intercept Form: $y = mx + b$, where the slope is *m* and the *y* intercept is *b*.

3. Point-Slope Form: $y - y_1 = m(x - x_1)$, where the slope is *m* and (x_1, y_1) is any point on the chosen line.

4. Two-Point Form: $\frac{y-y_1}{x-x_1} = \frac{y_2-y_1}{x_2-x_1}$, where (x_1, y_1) and (x_2, y_2) are any two distinct points on the chosen line. Note that the slope is given by $m = \frac{y_2-y_1}{x_2-x_1}$.

5. Intercept Form: $\frac{x}{x_1} + \frac{y}{y_1} = 1$, in which x_1 is the *x*-intercept and y_1 is the *y*-intercept.

These five ways to write linear equations are all useful in different circumstances. Depending on the given information, it may be easier to write one of the forms over another.

If $y = mx$, *y* is directly proportional to *x*. In this case, changing *x* by a factor changes *y* by that same factor. If $y = \frac{m}{x}$, *y* is inversely proportional to *x*. For example, if *x* is increased by a factor of 3, then *y* will be decreased by the same factor, 3.

Whether an expression or an equation, linear relationships describe the way two quantities change with respect to each other. The relationship is defined as linear because a line is produced if all the sets of corresponding values are graphed on a coordinate grid. When expressing the linear relationship as an equation, the equation is often written in the form $y = mx + b$ (slope-intercept form) where m and b are numerical values and *x* and *y* are variables (for example, $y = 5x + 10$). Given a linear equation and the value of either variable (*x* or *y*), the value of the other variable can be determined.

Suppose a teacher is grading a test containing 20 questions with 5 points given for each correct answer, adding a curve of 10 points to each test. This linear relationship can be expressed as the equation $y = 5x + 10$ where *x* represents the number of correct answers, and y represents the test score. To determine the score of a test with a given number of correct answers, the number of correct answers is substituted into the equation for *x* and evaluated. For example, for 10 correct answers, 10 is substituted for *x*: $y = 5(10) + 10 \rightarrow y = 60$. Therefore, 10 correct answers will result in a score of 60. The number of correct answers needed to obtain a certain score can also be determined. To determine the number of correct answers needed to score a 90, 90 is substituted for *y* in the equation (*y* represents the test score) and solved: $90 = 5x + 10 \rightarrow 80 = 5x \rightarrow 16 = x$. Therefore, 16 correct answers are needed to score a 90.

Linear relationships may be represented by a table of 2 corresponding values. Certain tables may determine the relationship between the values and predict other corresponding sets. Consider the table below, which displays the money in a checking account that charges a monthly fee:

Month	0	1	2	3	4
Balance	\$210	\$195	\$180	\$165	\$150

An examination of the values reveals that the account loses \$15 every month (the month increases by one and the balance decreases by 15). This information can be used to predict future values. To determine what the value will be in month 6, the pattern can be continued, and it can be concluded that the balance will be \$120. To determine which month the balance will be \$0, \$210 is divided by \$15 (since the balance decreases \$15 every month), resulting in month 14.

Similar to a table, a graph can display corresponding values of a linear relationship.

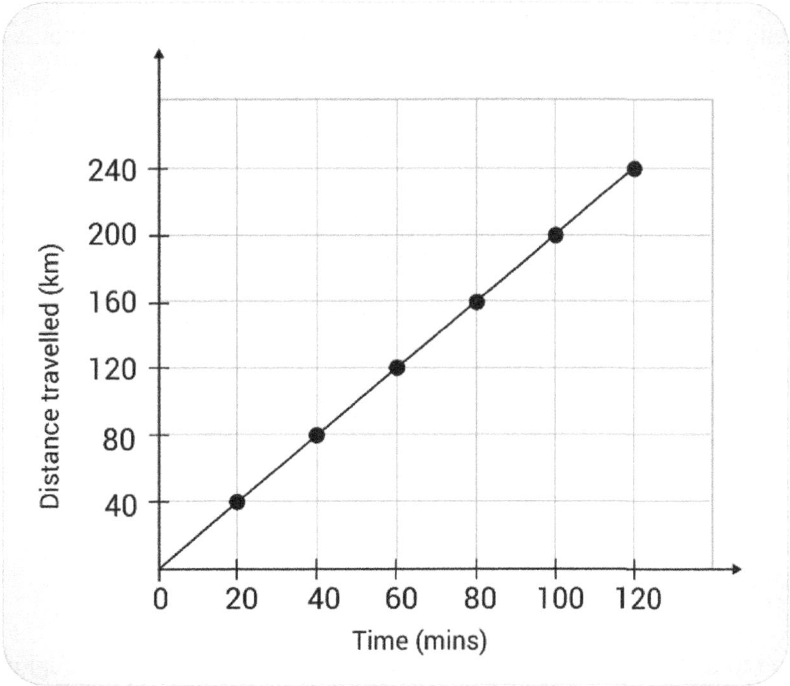

The graph above represents the relationship between distance traveled and time. To find the distance traveled in 80 minutes, the mark for 80 minutes is located at the bottom of the graph. By following this mark directly up on the graph, the corresponding point for 80 minutes is directly across from the 160 kilometer mark. This information indicates that the distance travelled in 80 minutes is 160 kilometers. To predict information not displayed on the graph, the way in which the variables change with respect to one another is determined. In this case, distance increases by 40 kilometers as time increases by 20 minutes. This information can be used to continue the data in the graph or convert the values to a table.

Solving Linear Equations

Solving equations in one variable is the process of isolating that variable on one side of the equation. The letters in an equation are variables as they stand for unknown quantities that you are trying to solve for. The numbers attached to the variables by multiplication are called coefficients. *X* is commonly used as a variable, though any letter can be used. For example, in $3x - 7 = 20$, the variable is $3x$, and it needs to be isolated. The numbers (also called **constants**) are -7 and 20. That means $3x$ needs to be on one side of the equals sign (either side is fine), and all the numbers need to be on the other side of the equals sign.

To accomplish this, the equation must be manipulated by performing opposite operations of what already exists. Remember that addition and subtraction are opposites and that multiplication and division are opposites. Any action taken to one side of the equation must be taken on the other side to maintain equality.

So, since the 7 is being subtracted, it can be moved to the right side of the equation by adding seven to both sides:

$$3x - 7 = 20$$

$$3x - 7 + 7 = 20 + 7$$

$$3x = 27$$

Now that the variable $3x$ is on one side and the constants (now combined into one constant) are on the other side, the 3 needs to be moved to the right side. 3 and x are being multiplied together, so 3 then needs to be divided from each side.

$$\frac{3x}{3} = \frac{27}{3}$$

$$x = 9$$

Now that x has been completely isolated, we know its value.

The solution is found to be $x = 9$. This solution can be checked for accuracy by plugging $x = 9$ in the original equation. After simplifying the equation, $20 = 20$ is found, which is a true statement:

$$3 \times 9 - 7 = 20$$

$$27 - 7 = 20$$

$$20 = 20$$

Equations that require solving for a variable (**algebraic equations**) come in many forms. Here are some more examples:

Sometimes, no number is attached to the variable:

$$x + 8 = 20$$

$$x + 8 - 8 = 20 - 8$$

$$x = 12$$

A fraction may be in the variable:

$$\frac{1}{2}z + 24 = 36$$

$$\frac{1}{2}z + 24 - 24 = 36 - 24$$

$$\frac{1}{2}z = 12$$

Now we multiply the variable by its inverse:

$$\frac{2}{1} \times \frac{1}{2} z = 12 \times \frac{2}{1}$$

$$z = 24$$

Some algebraic equations contain multiple instances of x:

$$14x + x - 4 = 3x + 2$$

All instances of x can be combined.

$$15x - 4 = 3x + 2$$

$$15x - 4 + 4 = 3x + 2 + 4$$

$$15x = 3x + 6$$

$$15x - 3x = 3x + 6 - 3x$$

$$12x = 6$$

$$\frac{12x}{12} = \frac{6}{12}$$

$$x = \frac{1}{2}$$

Linear Equations with One Solution, No Solution, or Infinitely Many Solutions

Although linear equations generally have one solution, this is not always the case. If there is no value for the variable that makes the statement true, there is no solution to the equation. Consider the equation $x + 3 = x - 1$. There is no value for x in which adding 3 to the value produces the same result as subtracting one from the value. Conversely, if any value for the variable makes a true statement, the equation has an infinite number of solutions. Consider the equation $3x + 6 = 3(x + 2)$. Any number substituted for x will result in a true statement (both sides of the equation are equal).

By manipulating equations like the two above, the variable of the equation will cancel out completely. If the remaining constants express a true statement (ex. $6 = 6$), then all real numbers are solutions to the equation. If the constants left express a false statement (ex. $3 = -1$), then no solution exists for the equation.

Solving Linear Equations with Coefficients

When solving a linear equation, the desired result requires determining a numerical value for the unknown variable. If given a linear equation involving addition, subtraction, multiplication, or division, working backwards isolates the variable. Addition and subtraction are inverse operations, as are multiplication and division. Therefore, they can be used to cancel each other out.

The first steps to solving linear equations are distributing, if necessary, and combining any like terms on the same side of the equation. Sides of an equation are separated by an **equal sign**. Next, the equation is manipulated to show the variable on one side. Whatever is done to one side of the equation must be done to the other side of the equation to remain equal. Inverse operations are then used to isolate the variable and undo the order of operations backwards. Addition and subtraction are undone, then multiplication and division are undone.

For example, solve $4(t-2)+2t-4=2(9-2t)$

Distributing: $4t-8+2t-4=18-4t$

Combining like terms: $6t-12=18-4t$

Adding $4t$ to each side to move the variable: $10t-12=18$

Adding 12 to each side to isolate the variable: $10t=30$

Dividing each side by 10 to isolate the variable: $t=3$

The answer can be checked by substituting the value for the variable into the original equation, ensuring that both sides calculate to be equal.

Linear equations and linear inequalities are both comparisons of two algebraic expressions. However, unlike equations in which the expressions are equal, linear inequalities compare expressions that may be unequal. Linear equations typically have one value for the variable that makes the statement true. Linear inequalities generally have an infinite number of values that make the statement true.

Linear inequalities express the relationship between unequal values. More specifically, they describe in what way the values are unequal. A value can be greater than (>), less than (<), greater than or equal to (≥), or less than or equal to (≤) another value. $5x+40>65$ is read as *five times a number added to forty is greater than sixty-five.*

When solving a linear inequality, the solution is the set of all numbers that make the statement true. The inequality $x+2\geq 6$ has a solution set of 4 and every number greater than 4 (4.01; 5; 12; 107; etc.). Adding 2 to 4 or any number greater than 4 results in a value that is greater than or equal to 6. Therefore, $x\geq 4$ is the solution set.

To algebraically solve a linear inequality, follow the same steps as those for solving a linear equation. The inequality symbol stays the same for all operations *except* when multiplying or dividing by a negative number. If multiplying or dividing by a negative number while solving an inequality, the relationship reverses (the sign flips). In other words, > switches to < and vice versa. Multiplying or dividing by a positive number does not change the relationship, so the sign stays the same. An example is shown below.

Solve $-2x-8\leq 22$

Add 8 to both sides: $-2x\leq 30$

Divide both sides by -2: $x\geq -15$

Solutions of a linear equation or a linear inequality are the values of the variable that make a statement true. In the case of a linear equation, the solution set (list of all possible solutions) typically consists of a single numerical value. To find the solution, the equation is solved by isolating the variable. For example, solving the equation $3x-7=-13$ produces the solution $x=-2$. The only value for *x* which produces a true statement is -2. This can be checked by substituting -2 into the original equation to check that both sides are equal. In this case, $3(-2) - 7 = -13 \rightarrow -13 = -13$; therefore, -2 is a solution.

Solving a linear inequality requires all values that make the statement true to be determined. For example, solving $3x - 7 \geq -13$ produces the solution $x \geq -2$. This means that -2 and any number greater than -2 produces a true statement. Solution sets for linear inequalities will often be displayed using a number line. If a value is included in the set (≥ or ≤), a shaded dot is placed on that value and an arrow extending in the direction of the solutions. For a variable > or ≥ a number, the arrow will point right on a number line, the direction where the numbers increase. If a variable is < or ≤ a number, the arrow will point left on a number line, which is the direction where the numbers decrease. If the value is not included in the set (> or <), an open (unshaded) circle on that value is used with an arrow in the appropriate direction.

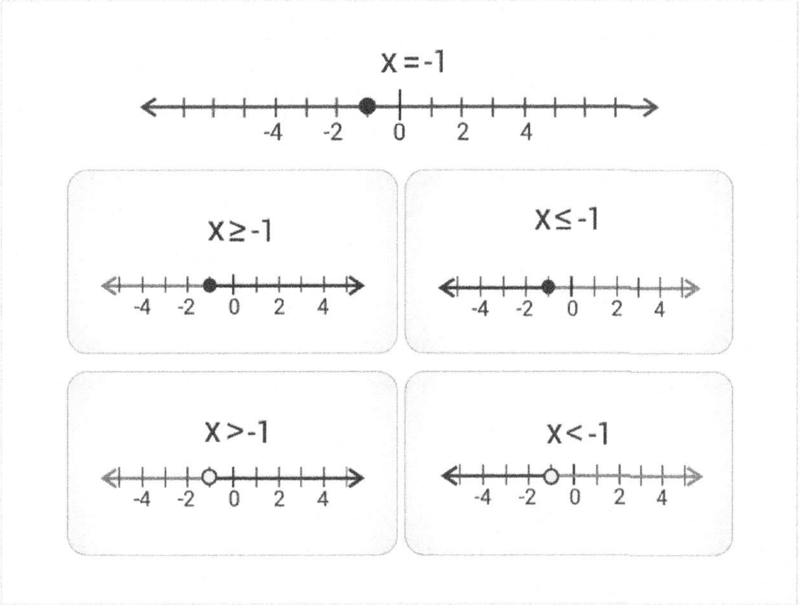

Similar to linear equations, a linear inequality may have a solution set consisting of all real numbers or can contain no solution. When solved algebraically, a linear inequality in which the variable cancels out and results in a true statement (ex. $7 \geq 2$) has a solution set of all real numbers. A linear inequality in which the variable cancels out and results in a false statement (ex. $7 \leq 2$) has no solution.

Imagine the following problem: The sum of a number and 5 is equal to -8 times the number.

To find this unknown number, a simple equation can be written to represent the problem. Key words such as difference, equal, and times are used to form the following equation with one variable: $n + 5 = -8n$. When solving for n, opposite operations are used. First, n is subtracted from $-8n$ across the equals sign, resulting in $5 = -9n$. Then, -9 is divided on both sides, leaving $n = -\frac{5}{9}$. This solution can be graphed on the number line with a dot as shown below:

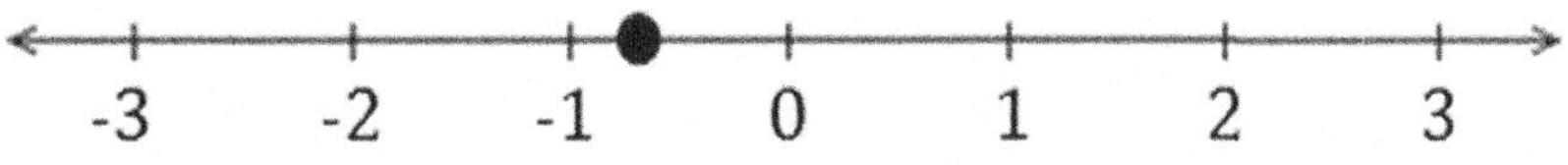

If the problem were changed to say, "The sum of a number and 5 is greater than -8 times the number," then an inequality would be used instead of an equation. Using key words again, *greater than* is represented by the symbol >. The inequality $n + 5 > -8n$ can be solved using the same techniques, resulting in $n < -\frac{5}{9}$. The only time solving an inequality differs from solving an equation is when a negative number is either multiplied by or divided by each side of the inequality. The sign must be switched in this case. For this example, the graph of the solution changes to the following graph because the solution represents all real numbers less than $-\frac{5}{9}$. Not included in this solution is $-\frac{5}{9}$ because it is a *less than* symbol, not *equal to*.

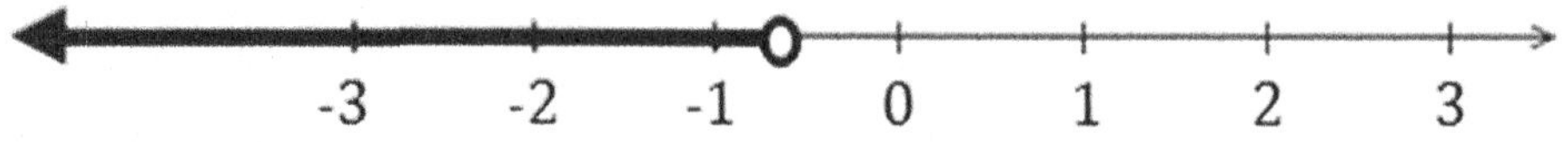

Analyzing and Solving Simultaneous Linear Equations

Understanding the Solutions to Systems of Two Linear Equations

A **system of equations** is a group of equations that have the same variables or unknowns. These equations can be linear, but they are not always so. Finding a solution to a system of equations means finding the values of the variables that satisfy each equation. For a linear system of two equations and two variables, there could be a single solution, no solution, or infinitely many solutions.

A single solution occurs when there is one value for x and y that satisfies the system. This would be shown on the graph where the lines cross at exactly one point. When there is no solution, the lines are parallel and do not ever cross. With infinitely many solutions, the equations may look different, but they are the same line. One equation will be a multiple of the other, and on the graph, they lie on top of each other.

One method for solving a system of equations is to graph both lines on a coordinate plane. If the lines intersect, the point of intersection is the solution to the system. Every point on a line represents an ordered pair that makes its equation true. The ordered pair represented by this point of intersection lies on both lines and therefore makes both equations true. This ordered pair should be checked by substituting its values into both of the original equations of the system. Note that given a system of equations and an ordered pair, the ordered pair can be determined to be a solution or not by checking it in both equations.

Again, the solution to a linear system is the point or points where the lines cross. The values of x and y represent the coordinates (x, y) where the lines intersect. Using the system $x + 3y = 10$ and $2x - 5y = 9$, the equations can be solved for y to put them in slope-intercept form, $y = mx + b$. These equations become $y = x + 2$ and $y = -\frac{3}{2}x + 4.5$. The slope is the coefficient of x, and the y-intercept is the constant value.

This system with the solution is shown below:

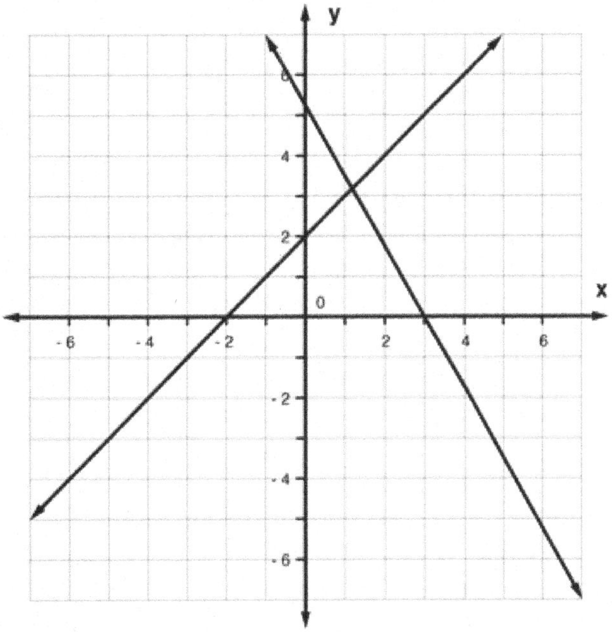

If, when graphed, the lines representing the equations of a system do not intersect, then the two lines are parallel to each other or they are the same exact line. Parallel lines extend in the same direction without ever meeting. A system consisting of parallel lines has no solution. If the equations for a system represent the same exact line, then every point on the line is a solution to the system. In this case, there would be an infinite number of solutions. A system consisting of intersecting lines is referred to as independent; a system consisting of parallel lines is referred to as inconsistent; and a system consisting of coinciding lines is referred to as dependent.

Parallel Lines	**Intersecting Lines**	**Coincident Lines**
Inconsistent	Independent	Dependent

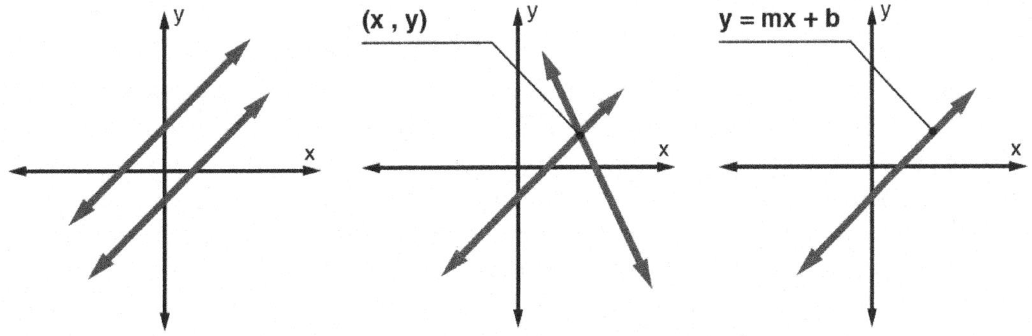

Systems of Equations with No Solution or an Infinite Number of Solutions
A system of equations can have one solution, no solution, or an infinite number of solutions. If, while solving a system algebraically, both variables cancel out, then the system has either no solution or has an infinite number of solutions. If the remaining constants result in a true statement (ex., $7 = 7$), then there is an infinite number of solutions. This would indicate coinciding lines. If the remaining constants result in a false statement, then there is no solution to the system. This would indicate parallel lines.

Solving Systems of Two Linear Equations in Two Variables Algebraically
A system of two linear equations in two variables is a set of equations that use the same variables (typically *x* and *y*). A solution to the system is an ordered pair that makes both equations true. One method for solving a system is by graphing. This method, however, is not always practical. Students may not have graph paper; or the solution may not consist of integers, making it difficult to identify the exact point of intersection on a graph. There are two methods for solving systems of equations algebraically: substitution and elimination. The method used will depend on the characteristics of the equations in the system.

Solving Systems of Equations with the Substitution Method
If one of the equations in a system has an isolated variable (*x*= or *y*=) or a variable that can be easily isolated, the substitution method can be used. This involves solving one equation for a variable and then plugging that solved equation into the other equation in the system. This equation can be solved for one variable, which can then be plugged in to either original equation and solved for the other variable. For example, $x - y = -2$ and $3x + 2y = 9$ can be solved using substitution. The first equation can be solved for x, where $x = -2 + y$.

Then it can be plugged into the other equation:

$$3(-2 + y) + 2y = 9$$

Solving for y yields:

$$-6 + 3y + 2y = 9$$

That shows that $y = 3$. If $y = 3$, then $x = 1$.

This solution can be checked by plugging in these values for the variables in each equation to see if it makes a true statement.

Here's another sample system: $x + 3y = 7; 2x - 4y = 24$. The first equation can easily be solved for *x*. By subtracting 3*y* on both sides, the resulting equation is $x = 7 - 3y$. When one equation is solved for a variable, the expression that it is equal can be substituted into the other equation. For this example, $(7 - 3y)$ would be substituted for *x* into the second equation as follows: $2(7 - 3y) - 4y = 24$. Solving this equation results in $y = -1$. Once the value for one variable is known, this value should be substituted into either of the original equations to determine the value of the other variable. For the example, –1 would be substituted for *y* in either of the original equations. Substituting into the first equation results in $x + 3(-1) = 7$, and solving this equation yields $x = 10$. The solution to a system is an ordered pair, so the solution to the example is written as (10, –1). The solution can be checked by substituting it into both equations of the system to ensure it results in two true statements.

Solving Systems of Equations with the Elimination Method

The elimination method for solving a system of equations involves canceling out (or eliminating) one of the variables. This method is typically used when both equations of a system are written in standard form ($Ax + By = C$).

Let's return to the system of equations we graphed:

$$x + 3y = 10 \text{ and } 2x - 5y = 9$$

Immediately adding these equations does not eliminate a variable, but it is possible to change the first equation by multiplying the whole equation by -2. This changes the first equation to

$$-2x - 6y = -20$$

The equations can be then added to obtain $-11y = -11$. Solving for y yields $y = 1$. To find the rest of the solution, 1 can be substituted in for y in either original equation to find the value of $x = 7$. The solution to the system is (7, 1) because it makes both equations true, and it is the point in which the lines intersect. If the system is **dependent**—having infinitely many solutions—then both variables will cancel out when the elimination method is used, resulting in an equation that is true for many values of *x* and *y*. Since the system is dependent, both equations can be simplified to the same equation or line.

Another example is $2x + 3y = 12; 5x - y = 13$. To perform the elimination method, the equations in the system should be arranged vertically to be added together and then one or both of the equations should be multiplied so that one variable will be eliminated when the two are added. Opposites will cancel each other when added together. For example, 8*x* and -8*x* will cancel each other when added. For the example above, writing the system vertically helps identify that the bottom equation should be multiplied by 3 to eliminate the variable *y*.

$$2x + 3y = 12 \quad \rightarrow \quad 2x + 3y = 12$$

$$3(5x - y = 13) \quad \rightarrow \quad 15x - 3y = 39$$

Adding the two equations together vertically results in $17x = 51$. Solving yields $x = 3$. Once the value for one variable is known, it can be substituted into either of the original equations to determine the value of the other variable. Once this is obtained, the solution can be written as an ordered pair (*x*, *y*) and checked in both equations of the system. In this example, the solution is (3, 2).

Solving Problems Leading to Two Linear Equations in Two Variables

As mentioned, a system of two linear equations in two variables is a set of equations that use the same variables, usually *x* and *y*. Here's a sample problem:

> An Internet provider charges an installation fee and a monthly charge. It advertises that two months of its offering costs \$100 and six months costs \$200. Find the monthly charge and the installation fee.

The two unknown quantities (variables) are the monthly charge and the installation fee. There are two different statements given relating the variables: two months added to the installation fee is \$100; and six months added to the installation fee is \$200. Using the variable x as the monthly charge and y as the installation fee, the statements can be written as the following:

$$2x + y = 100$$

$$6x + y = 200$$

These two equations taken together form a system modeling the given scenario.

Functions

Defining, Evaluating, and Comparing Functions

Functions

First, it's important to understand the definition of a **relation**. Given two variables, *x* and *y*, which stand for unknown numbers, a **relation** between *x* and *y* is an object that splits all of the pairs (*x*, *y*) into those for which the relation is true and those for which it is false. For example, consider the relation of $x^2 = y^2$. This relationship is true for the pair (1, 1) and for the pair (-2, 2), but false for (2, 3). Another example of a relation is $x \leq y$. This is true whenever *x* is less than or equal to *y*.

A **function** is a special kind of relation where, for each value of *x*, there is only a single value of *y* that satisfies the relation. So, $x^2 = y^2$ is *not* a function because in this case, if *x* is 1, *y* can be either 1 or -1: the pair (1, 1) and (1, -1) both satisfy the relation. More generally, for this relation, any pair of the form $(a, \pm a)$ will satisfy it. On the other hand, consider the following relation: $y = x^2 + 1$. This is a function because for each value of *x*, there is a unique value of *y* that satisfies the relation. Notice, however, there are multiple values of *x* that give us the same value of *y*. This is perfectly acceptable for a function. Therefore, *y* is a function of *x*.

To determine if a relation is a function, check to see if every *x* value has a unique corresponding *y* value.

The **domain** of a function consists of all inputs, and the **range** consists of all outputs. Graphing the ordered pairs of a linear function produces a straight line. An example of a function would be $f(x) = 4x + 4$, read "*f* of *x* is equal to four times *x* plus four." In this example, the input would be *x* and the output would be f(*x*). Ordered pairs would be represented as (*x*, f(*x*)). To find the output for an input value of 3, 3 would be substituted for *x* into the function as follows: $f(3) = 4(3) + 4$, resulting in $f(3) = 16$. Therefore, the ordered pair $(3, f(3)) = (3, 16)$. Note f(*x*) is a function of *x* denoted by *f*. Functions of *x* could be named *g*(*x*), read "*g* of *x*"; *p*(*x*), read "*p* of *x*"; etc.

A linear function could also be written in the form of an equation with two variables. Typically, the variable *x* represents the inputs and the variable *y* represents the outputs. The variable *x* is considered the independent variable and *y* the dependent variable. The above function would be written as $y = 4x + 4$. Ordered pairs are written in the form (*x*, *y*).

Comparing Properties of Functions Represented in Different Ways

As mentioned, a **function** is defined as a relationship between inputs and outputs where there is only one output value for a given input. As an example, the following function is in function notation:

$$f(x) = 3x - 4$$

The $f(x)$ represents the output value for an input of x. If $x = 2$, the equation becomes:

$$f(2) = 3(2) - 4$$

$$6 - 4 = 2$$

The input of 2 yields an output of 2, forming the ordered pair $(2, 2)$. The following set of ordered pairs corresponds to the given function: $(2, 2), (0, -4), (-2, -10)$. Again, the set of all possible inputs of a function is its **domain**, and all possible outputs is called the **range**. By definition, each member of the domain is paired with only one member of the range.

Functions can also be defined recursively. In this form, they are not defined explicitly in terms of variables. Instead, they are defined using previously-evaluated function outputs, starting with either $f(0)$ or $f(1)$. An example of a recursively-defined function is:

$$f(1) = 2$$

$$f(n) = 2f(n - 1) + 2n$$

$$n > 1$$

The domain of this function is the set of all integers.

A function $f(x)$ is a mathematical object which takes one number, *x*, as an input and gives a number in return. The input is called the **independent variable**. If the variable is set equal to the output, as in $y = f(x)$, then this is called the **dependent variable**. To indicate the dependent value a function, y, gives for a specific independent variable, x, the notation y = $f(x)$ is used.

The **domain** of a function is the set of values that the independent variable is allowed to take. Unless otherwise specified, the domain is any value for which the function is well defined. The **range** of the function is the set of possible outputs for the function.

In many cases, a function can be defined by giving an equation. For instance, $f(x) = x^2$ indicates that given a value for *x*, the output of *f* is found by squaring *x*.

Not all equations in *x* and *y* can be written in the form $y = f(x)$. An equation can be written in such a form if it satisfies the **vertical line test**: no vertical line meets the graph of the equation at more than a single point. In this case, *y* is said to be a *function of x*. If a vertical line meets the graph in two places, then this equation cannot be written in the form $y = f(x)$.

The graph of a function $f(x)$ is the graph of the equation $y = f(x)$. Thus, it is the set of all pairs (x, y) where $y = f(x)$. In other words, it is all pairs $(x, f(x))$. The *x*-intercepts are called the **zeros** of the function. The *y*-intercept is given by $f(0)$.

If, for a given function f, the only way to get $f(a) = f(b)$ is for $a = b$, then f is *one-to-one*. Often, even if a function is not one-to-one on its entire domain, it is one-to-one by considering a restricted portion of the domain.

A function $f(x) = k$ for some number *k* is called a **constant function**. The graph of a constant function is a horizontal line.

The function $f(x) = x$ is called the **identity function**. The graph of the identity function is the diagonal line pointing to the upper right at 45 degrees, $y = x$.

A function is called **monotone** if it is either always increasing or always decreasing. For example, the functions $f(x) = 3x$ and $f(x) = -x^5$ are monotone.

An **even function** looks the same when flipped over the *y*-axis: $f(x) = f(-x)$. The following image shows a graphic representation of an even function.

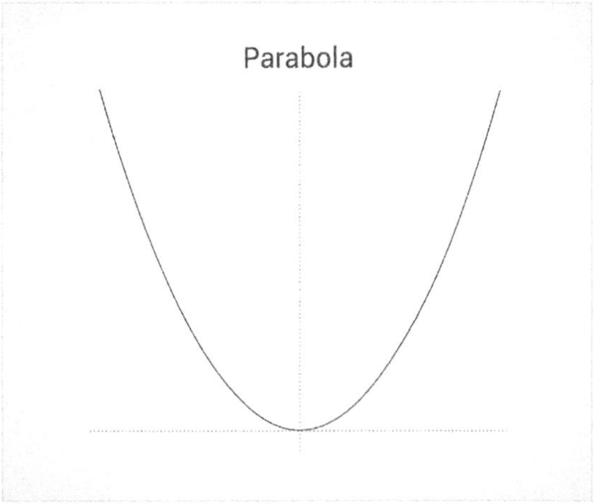

An **odd function** looks the same when flipped over the *y*-axis and then flipped over the *x*-axis:

$$f(x) = -f(-x)$$

The following image shows an example of an odd function.

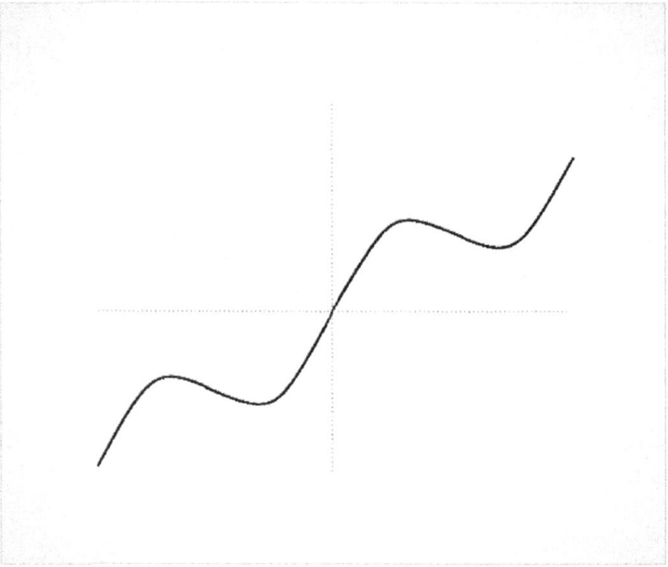

A function is called **linear** if it can take the form of the equation $f(x) = ax + b$, or $y = ax + b$, for any two numbers a and b. A linear equation forms a straight line when graphed on the coordinate plane. An example of a linear function is shown below on the graph.

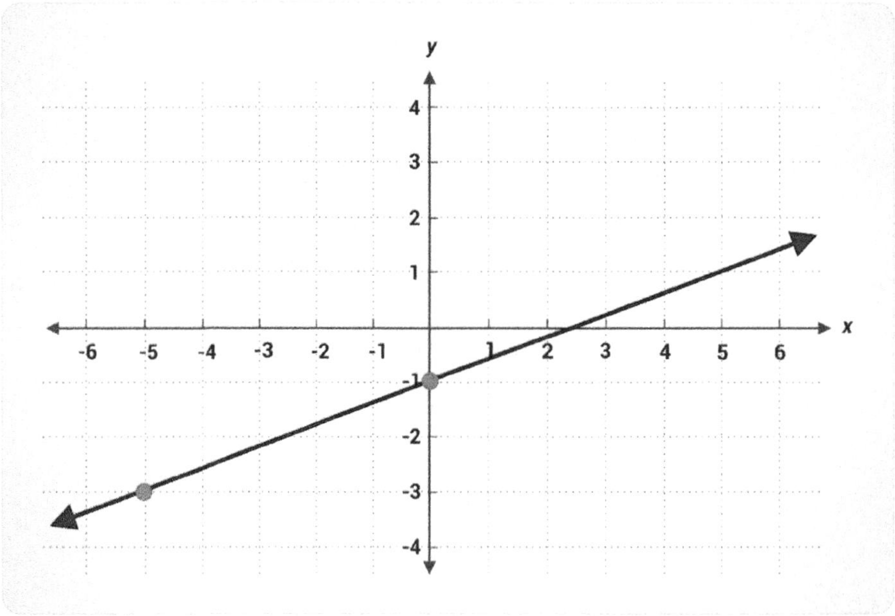

This is a graph of the following function:

$$y = \frac{2}{5}x - 1$$

A table of values that satisfies this function is shown below.

x	y
-5	-3
0	-1
5	1
10	3

These points can be found on the graph using the form (x,y).

Given two ordered pairs for a **linear** function, (x_1, y_1) and (x_2, y_2), it is possible to determine the rate of change between the variables (slope of the line). To calculate the slope of the line, m, the values for the ordered pairs should be substituted into the formula:

$$m = \frac{y_2 - y_1}{x_2 - x_1}$$

The expression is substituted to obtain a whole number or fraction for the slope. Once the slope is calculated, the slope and either of the ordered pairs should be substituted into the point-slope form to obtain the equation of the line.

Lastly, the same function can be represented in different ways, including a diagram, a table, a graph, and a set of numbers simultaneously.

Here is such an example:

Multiple Representations of a Function

Mapping

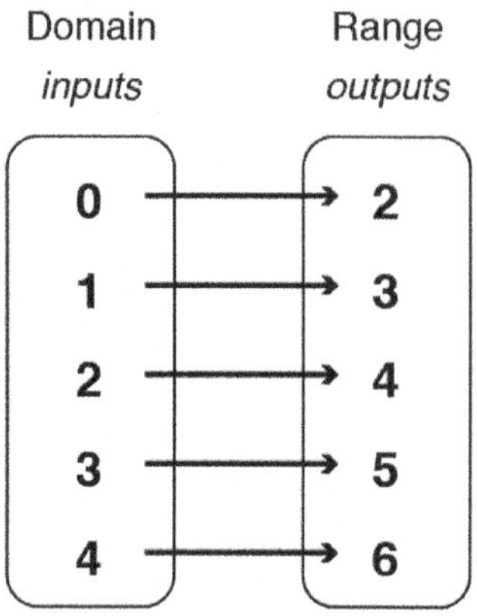

Table

x	y
0	2
1	3
2	4
3	5
4	6

Graph

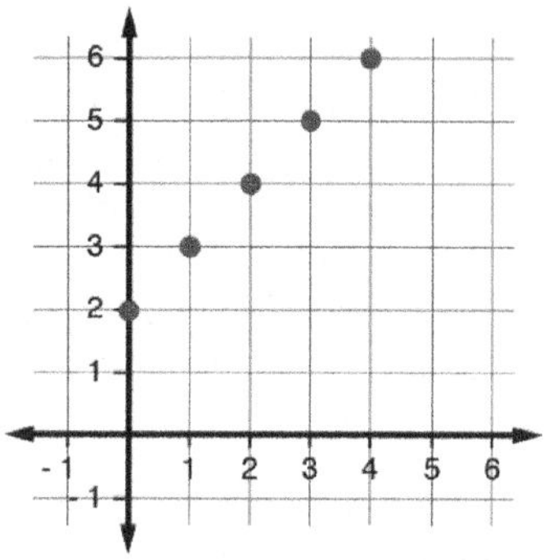

Ordered Pairs

{(0,2),(1,3),(2,4),(3,5),(4,6)}

It is important to be able to recognize and manipulate functions in all of their forms.

Linear and Non-Linear Functions

Three common functions used to model different relationships between quantities are linear, quadratic, and exponential functions. **Linear functions** are the simplest of the three, and the independent variable x has an exponent of 1. Written in the most common form, $y = mx + b$, the coefficient of x tells how fast the function grows at a constant rate, and the b-value tells the starting point. A **quadratic function** has an exponent of 2 on the independent variable x. Standard form for this type of function is $y = ax^2 + bx + c$, and the graph is a parabola. These type functions grow at a changing rate. **An exponential function** has an independent variable in the exponent $y = ab^x$. The graph of these types of functions is described as **growth** or **decay**, based on whether the **base**, b, is greater than or less than 1. These functions are different from quadratic functions because the base stays constant. A common base is base e.

The following three functions model a linear, quadratic, and exponential function respectively: $y = 2x$, $y = x^2$, and $y = 2^x$. Their graphs are shown below. The first graph, modeling the linear function, shows that the growth is constant over each interval. With a horizontal change of 1, the vertical change is 2. It models a constant positive growth. The second graph shows the quadratic function, which is a curve that is symmetric across the y-axis. The growth is not constant, but the change is mirrored over the axis. The last graph models the exponential function, where the horizontal change of 1 yields a vertical change that increases more and more. The exponential graph gets very close to the x-axis, but never touches it, meaning there is an asymptote there. The y-value can never be zero because the base of 2 can never be raised to an input value that yields an output of zero.

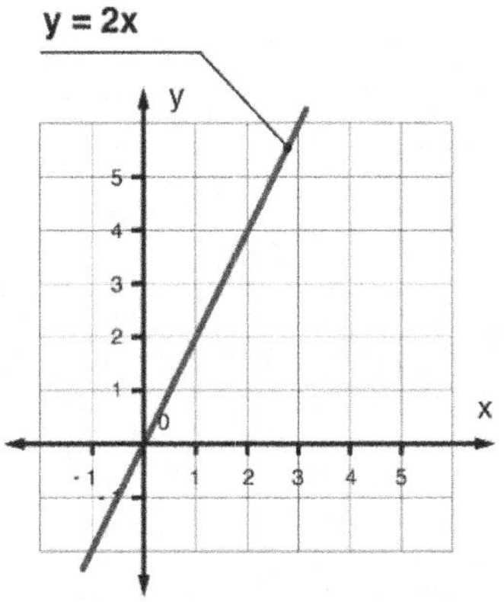

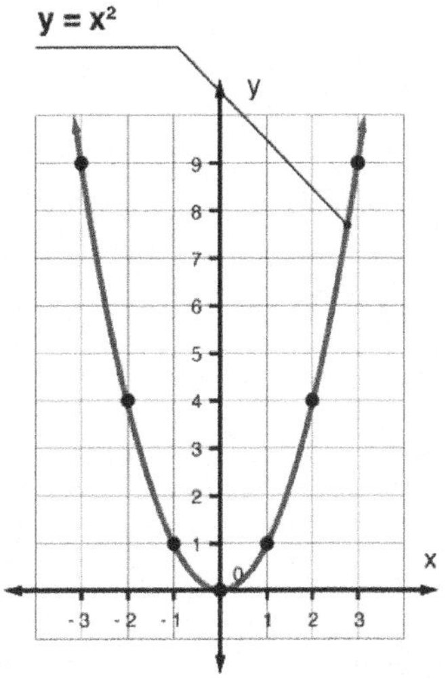

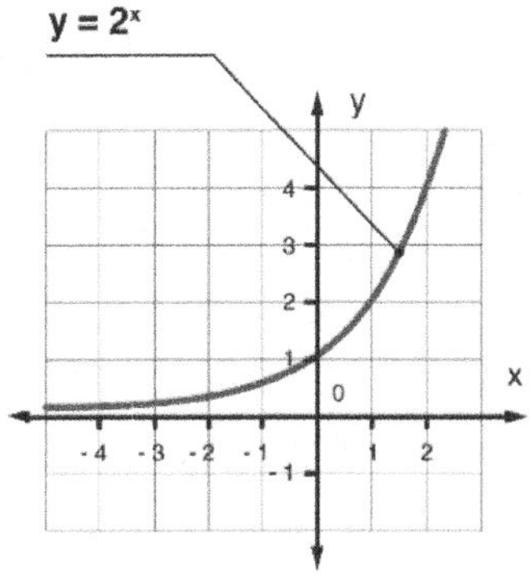

The three tables below show specific values for three types of functions. The third column in each table shows the change in the y-values for each interval. The first table shows a constant change of 2 for each equal interval, which matches the slope in the equation $y = 2x$. The second table shows an increasing change, but it also has a pattern. The increase is changing by 2 more each time, so the change is quadratic. The third table shows the change as factors of the base, 2. It shows a continuing pattern of factors of the base.

$y = 2x$		
x	y	Δy
1	2	
2	4	2
3	6	2
4	8	2
5	10	2

$y = x^2$		
x	y	Δy
1	1	
2	4	3
3	9	5
4	16	7
5	25	9

$y = 2^x$		
x	y	Δy
1	2	
2	4	2
3	8	4
4	16	8
5	32	16

Given a table of values, the type of function can be determined by observing the change in y over equal intervals. For example, the tables below model two functions. The changes in interval for the x-values is 1 for both tables. For the first table, the y-values increase by 5 for each interval. Since the change is constant, the situation can be described as a linear function. The equation would be $y = 5x + 3$. For the second table, the change for y is 5, 20, 100, and 500, respectively. The increases are multiples of 5, meaning the situation can be modeled by an exponential function. The equation $y = 5^x + 3$ models this situation.

x	y
0	3
1	8
2	13
3	18
4	23

x	y
0	3
1	8
2	28
3	128
4	628

Quadratic equations can be used to model real-world area problems. For example, a farmer may have a rectangular field that he needs to sow with seed. The field has length $x + 8$ and width $2x$. The formula for area should be used: $A = lw$. Therefore:

$$A = (x + 8) \times 2x = 2x^2 + 16x$$

The possible values for the length and width can be shown in a table, with input x and output A. If the equation was graphed, the possible area values can be seen on the y-axis for given x-values.

Exponential growth and decay can be found in real-world situations. For example, if a piece of notebook paper is folded 25 times, the thickness of the paper can be found. To model this situation, a table can be used. The initial point is one-fold, which yields a thickness of 2 papers. For the second fold, the thickness is 4. Since the thickness doubles each time, the table below shows the thickness for the next few folds. Notice the thickness changes by the same factor each time. Since this change for a constant interval of folds is a factor of 2, the function is exponential. The equation for this is $y = 2^x$. For twenty-five folds, the thickness would be 33,554,432 papers.

x **(folds)**	y **(paper thickness)**
0	1
1	2
2	4
3	8
4	16
5	32

One exponential formula that is commonly used is the **interest formula**:

$$A = Pe^{rt}$$

In this formula, interest is compounded continuously. A is the value of the investment after the time, t, in years. P is the initial amount of the investment, r is the interest rate, and e is the constant equal to approximately 2.718. Given an initial amount of $200 and a time of 3 years, if interest is compounded continuously at a rate of 6%, the total investment value can be found by plugging each value into the formula. The invested value at the end is \$239.44. In more complex problems, the final investment may be given, and the rate may be the unknown. In this case, the formula becomes $239.44 = 200e^{r3}$. Solving for r requires isolating the exponential expression on one side by dividing by 200, yielding the equation $1.20 = e^{r3}$. Taking the natural log of both sides results in $\ln(1.2) = r3$. Using a calculator to evaluate the logarithmic expression, $r = 0.06 = 6\%$.

When working with logarithms and exponential expressions, it is important to remember the relationship between the two. In general, the logarithmic form is $y = log_b x$ for an exponential form $b^y = x$. Logarithms and exponential functions are inverses of each other.

Using Functions to Model Relationships Between Quantities

Constructing Functions to Model Linear Relationships Between Two Quantities

A **linear function** of the form $y = mx + b$ has two important quantities: *m* and *b*. The variable *m* represents the slope of the line, and the *b* indicates the *y*-intercept of the line. Slope is equivalent to the rate of change or the amount that *y* changes for a change in *x*. A positive value of *m* indicates a positive rate of change, while a negative value of *m* indicates a negative rate of change. The *y*-intercept, *b*, indicates the *y*-value, or the quantity of *y,* when *x* is equal to 0.

The two quantities, *m* and *b,* are especially meaningful when the linear equation models (or represents) an actual real-life situation. In many of these real-world applications, the *y*-intercept represents the initial amount of the quantity *y* because the *x*-variable would never be a negative quantity. For example, *x* often indicates time, and in many of these scenarios, the *x*-variable is never negative because time would not be negative.

Consider the following example. These two equations represent the cost, *C*, of gigabytes of data on a phone plan, *g*, at two different cellular phone companies:

$$C = 9g$$

$$C = 6g + 18$$

The first equation represents a scenario that shows the cost per gigabyte of data is \$9. In this equation, *g* varies directly with *y*. There is no *y*-intercept, which means that there is no initial cost for a data plan with that company. The rate of change is 9, which is price per gigabyte of data. The second equation represents a scenario that has both an initial cost and a cost per gigabyte of data. The slope, which is 6, means that each gigabyte is \$6. The *y*-intercept of 18 shows that there is an initial cost of a data plan with that company. Therefore, it makes sense to use the first company at \$9 a gigabyte when only purchasing a small number of gigabytes of data. However, any large data needs would be cheaper by going with the second company because eventually that initial cost will be negligible. The exact tipping point can be determined by setting the equations equal to one another and solving for *g:*

$$9g = 6g + 18$$

$$3g = 18$$

$$g = 6$$

Therefore, both companies charge the same when 6 gigabytes of data are purchased. When fewer than 6 gigabytes are purchased, the first company is a better value because it doesn't charge the initial cost, and when more than 6 gigabytes of data are purchased, the second company is a better value because the rate per gigabyte is less.

Let's try building another function in the context of a situation. For example, the relationship between the money paid for a gym membership and the months that someone has been a member can be described through a function. If the one-time membership fee is \$40 and the monthly fee is \$30, then the function can be written $f(x) = 30x + 40$. The x-value represents the number of months the person has been part of the gym, while the output is the total money paid for the membership. The table below shows this relationship. It is a representation of the function because the initial cost is \$40 and the cost increases each month by \$30.

x (months)	y (money paid to gym)
0	40
1	70
2	100
3	130

Analyzing Graphs of Functions and Graphing Functions

To graph relations and functions, the Cartesian plane is used. This means to think of the plane as being given a grid of squares, with one direction being the *x*-axis and the other direction the *y*-axis. Generally, the independent variable is placed along the horizontal axis, and the dependent variable is placed along the vertical axis. Any point on the plane can be specified by saying how far to go along the *x*-axis and how far along the *y*-axis with a pair of numbers (x, y). Specific values for these pairs can be given names such as $C = (-1, 3)$. Negative values mean to move left or down; positive values mean to move right or up. The point where the axes cross one another is called the **origin.** The origin has coordinates $(0, 0)$ and is usually called *O* when given a specific label. An illustration of the Cartesian plane, along with graphs of $(2, 1)$ and $(-1, -1)$, are below.

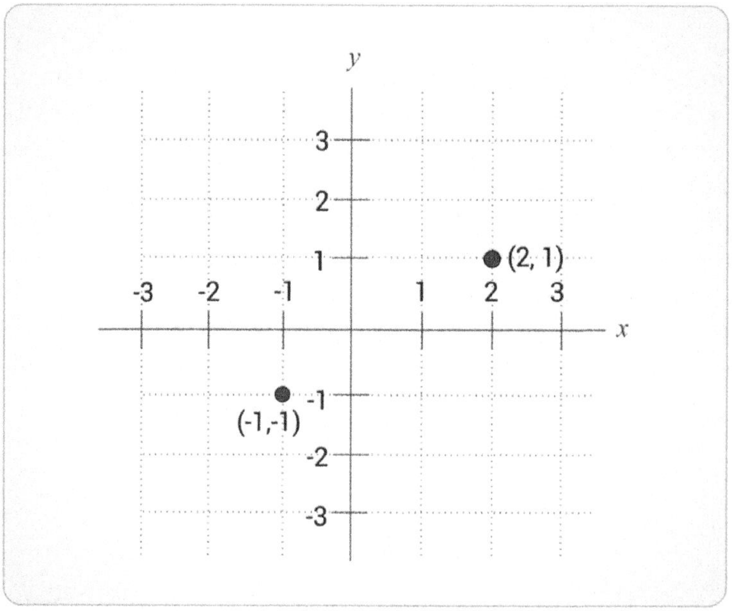

Relations also can be graphed by marking each point whose coordinates satisfy the relation. If the relation is a function, then there is only one value of *y* for any given value of x. This leads to the **vertical line test**: if a relation is graphed, then the relation is a function if any possible vertical line drawn anywhere along the graph would only touch the graph of the relation in no more than one place. Conversely, when graphing a function, then any possible vertical line drawn will not touch the graph of the function at any point or will touch the function at just one point. This test is made from the definition of a function, where each x-value must be mapped to one and only one y-value.

Graphing Linear Functions

When graphing a linear function, note that the ratio of the change of the *y* coordinate to the change in the *x* coordinate is constant between any two points on the resulting line, no matter which two points are chosen. In other words, in a pair of points on a line, (x_1, y_1) and (x_2, y_2), with $x_1 \neq x_2$ so that the two points are distinct, then the ratio $\frac{y_2-y_1}{x_2-x_1}$ will be the same, regardless of which particular pair of points are chosen. This ratio, $\frac{y_2-y_1}{x_2-x_1}$, is called the *slope* of the line and is frequently denoted with the letter m. If slope *m* is positive, then the line goes upward when moving to the right, while if slope *m* is negative, then the line goes downward when moving to the right. If the slope is 0, then the line is called horizontal, and the y coordinate is constant along the entire line. In lines where the x coordinate is constant along the entire line, y is not actually a function of x. For such lines, the slope is not defined. These lines are called vertical lines.

Linear functions may take forms other than $y = ax + b$. The most common forms of linear equations are explained below:

1. Standard Form: $Ax + By = C$, in which the slope is given by $m = \frac{-A}{B}$, and the *y*-intercept is given by $\frac{C}{B}$.

2. Slope-Intercept Form: $y = mx + b$, where the slope is *m* and the *y* intercept is *b*.

3. Point-Slope Form: $y - y_1 = m(x - x_1)$, where the slope is *m* and (x_1, y_1) is any point on the chosen line.

4. Two-Point Form: $\frac{y-y_1}{x-x_1} = \frac{y_2-y_1}{x_2-x_1}$, where (x_1, y_1) and (x_2, y_2) are any two distinct points on the chosen line. Note that the slope is given by $m = \frac{y_2-y_1}{x_2-x_1}$.

5. Intercept Form: $\frac{x}{x_1} + \frac{y}{y_1} = 1$, in which x_1 is the *x*-intercept and y_1 is the *y*-intercept.

These five ways to write linear equations are all useful in different circumstances. Depending on the given information, it may be easier to write one of the forms over another.

If $y = mx$, *y* is directly proportional to *x*. In this case, changing *x* by a factor changes *y* by that same factor. If $y = \frac{m}{x}$, *y* is inversely proportional to *x*. For example, if *x* is increased by a factor of 3, then *y* will be decreased by the same factor, 3.

The **midpoint** between two points, (x_1, y_1) and (x_2, y_2), is given by taking the average of the *x* coordinates and the average of the *y* coordinates:

$$\left(\frac{x_1 + x_2}{2}, \frac{y_1 + y_2}{2}\right)$$

The **distance** between two points, (x_1, y_1) and (x_2, y_2), is given by the **Pythagorean formula:**

$$\sqrt{(x_2 - x_1)^2 + (y_2 - y_1)^2}$$

To find the perpendicular distance between a line $Ax + By = C$ and a point (x_1, y_1) not on the line, we need to use the formula

$$\frac{|Ax_1 + By_1 + C|}{\sqrt{A^2 + B^2}}$$

Graphing a Quadratic Function

The standard form of a quadratic function is:

$$y = ax^2 + bx + c$$

The graph of a quadratic function is a u-shaped (or upside-down u) curve, called a parabola, which is symmetric about a vertical line (axis of symmetry). To graph a parabola, its vertex (high or low point for the curve) and at least two points on each side of the axis of symmetry need to be determined.

Given a quadratic function in standard form, the axis of symmetry for its graph is the line $x = -\frac{b}{2a}$. The vertex for the parabola has an x-coordinate of $-\frac{b}{2a}$. To find the y-coordinate for the vertex, the calculated x-coordinate needs to be substituted. To complete the graph, two different x-values need to be selected and substituted into the quadratic function to obtain the corresponding y-values. This will give two points on the parabola. These two points and the axis of symmetry are used to determine the two points corresponding to these. The corresponding points are the same distance from the axis of symmetry (on the other side) and contain the same y-coordinate. Plotting the vertex and four other points on the parabola allows for constructing the curve.

Quadratic Function

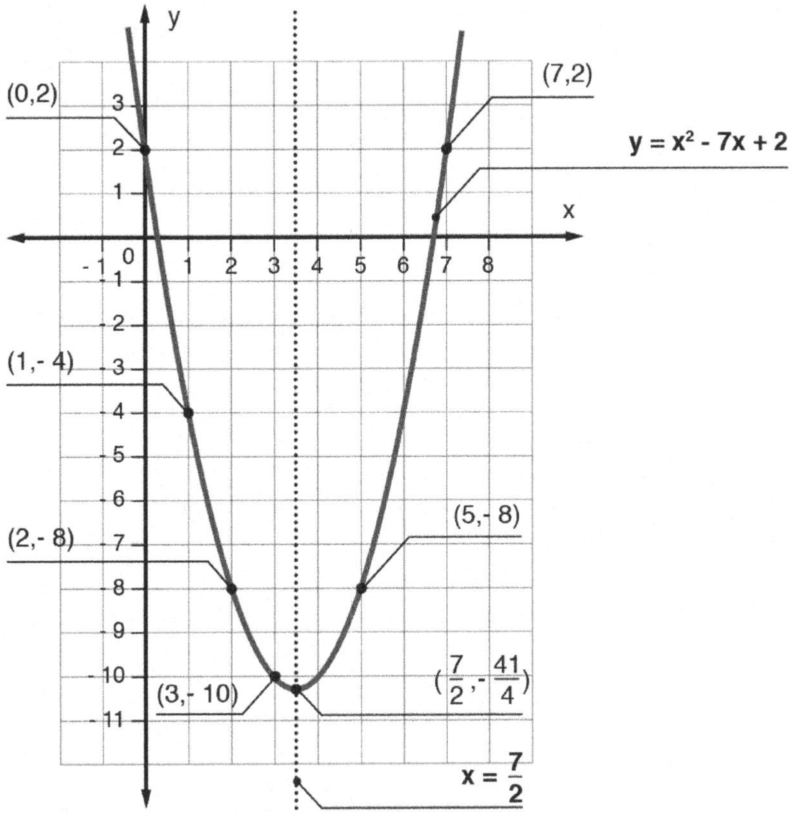

Graphing an Exponential Function

Exponential functions have a general form of $y = a \times b^x$. The graph of an exponential function is a curve that slopes upward or downward from left to right. The graph approaches a line, called an asymptote, as *x* or *y* increases or decreases. To graph the curve for an exponential function, *x*-values are selected and then substituted into the function to obtain the corresponding *y*-values. A general rule of thumb is to select three negative values, zero, and three positive values. Plotting the seven points on the graph for an exponential function should allow for constructing a smooth curve through them.

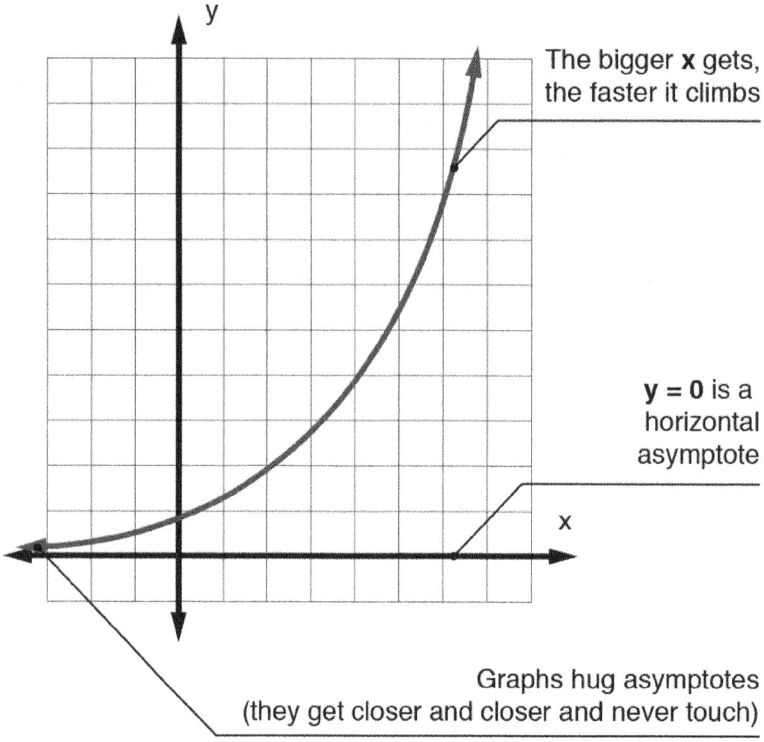

Note that **rate of change** for any line calculates the steepness of the line over a given interval. Rate of change is also known as the **slope** or rise/run. The rates of change for nonlinear functions vary depending on the interval being used for the function. The rate of change over one interval may be zero, while the next interval may have a positive rate of change. The equation plotted on the graph below, $y = x^2$, is a quadratic function and non-linear. The average rate of change from points $(0, 0)$ to $(1, 1)$ is 1 because the vertical change is 1 over the horizontal change of 1. For the next interval, $(1, 1)$ to $(2, 4)$, the average rate of change is 3 because the slope is $\frac{3}{1}$.

You can see that here:

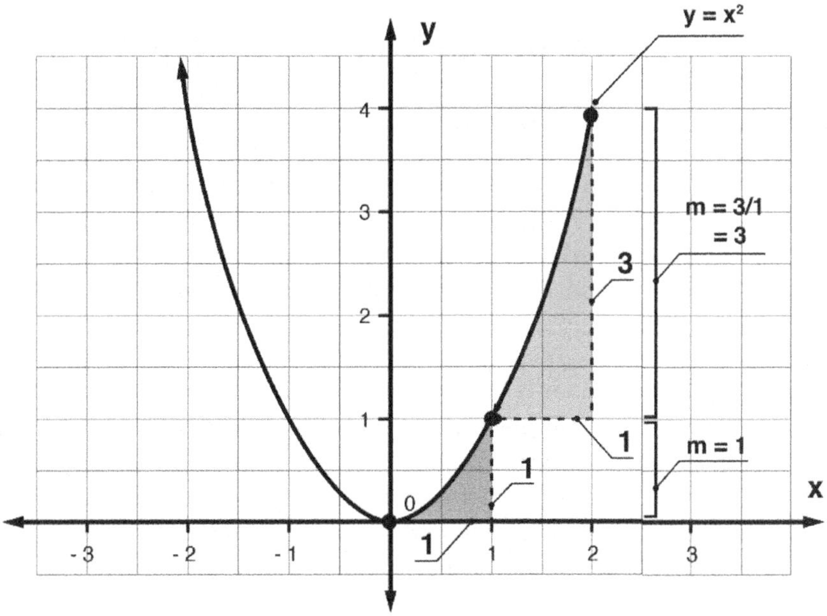

As mentioned previously, the rate of change for a linear function is constant and can be determined based on a few representations. One method is to place the equation in slope-intercept form:

$$y = mx + b$$

Thus, m is the slope, and b is the y-intercept. In the graph below, the equation is $y = x + 1$, where the slope is 1 and the y-intercept is 1. For every vertical change of 1 unit, there is a horizontal change of 1 unit. The x-intercept is -1, which is the point where the line crosses the x-axis.

Here's an example:

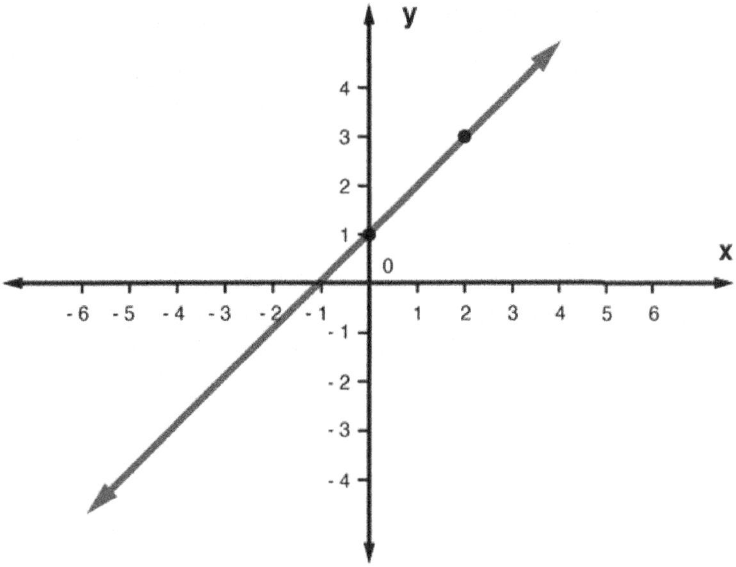

Geometry

Congruence and Similarity

Rotations, Reflections, Translations, and Dilations

A **transformation** occurs when a shape is altered in the plane where it exists. There are three major types of transformation: translations, reflections, and rotations. A **translation** consists of shifting a shape in one direction. A **reflection** results when a shape is transformed over a line to its mirror image. Finally, a **rotation** occurs when a shape moves in a circular motion around a specified point. The object can be turned clockwise or counterclockwise and, if rotated 360 degrees, returns to its original location.

The three major types of transformations preserve distance and angle measurement. The shapes stay the same, but they are moved to another place in the plane. Therefore, the distance between any two points on the shape doesn't change. Also, any original angle measure between two line segments doesn't change. However, there are transformations that don't preserve distance and angle measurements, including those that don't preserve the original shape. For example, transformations that involve stretching and shrinking shapes don't preserve distance and angle measures. In these cases, the input variables are multiplied by either a number greater than one (**stretch**) or less than one (**shrink**).

A **rigid motion** is a transformation that preserves distance and length. Every line segment in the resulting image is congruent to the corresponding line segment in the pre-image. Congruence between two figures means a series of transformations (or a rigid motion) can be defined that maps one of the figures onto the other. Basically, two figures are congruent if they have the same shape and size.

Translation means that all points in the figure are moved in the same direction by the same distance. In other words, the figure is slid in some fixed direction. Of course, while the entire figure is slid by the same distance, this does not change any of the measurements of the figures involved. The result will have the same distances and angles as the original figure.

In terms of Cartesian coordinates, a translation means a shift of each of the original points (x, y) by a fixed amount in the *x* and *y* directions, to become $(x + a, y + b)$.

For reflection to occur, a line in the plane is specified, called the **line of reflection**. Then, take each point and flip it over the line so that it is the same distance from the line but on the opposite side of it. This does not change any of the distances or angles involved, but it does reverse the order in which everything appears.

To reflect something over the *x*-axis, the points (x, y) are sent to $(x, -y)$. To reflect something over the *y*-axis, the points (x, y) are sent to the points $(-x, y)$. Flipping over other lines is not something easy to express in Cartesian coordinates. However, by drawing the figure and the line of reflection, the distance to the line and the original points can be used to find the reflected figure.

Example: Reflect this triangle with vertices (-1, 0), (2, 1), and (2, 0) over the *y*-axis. The pre-image is shown below.

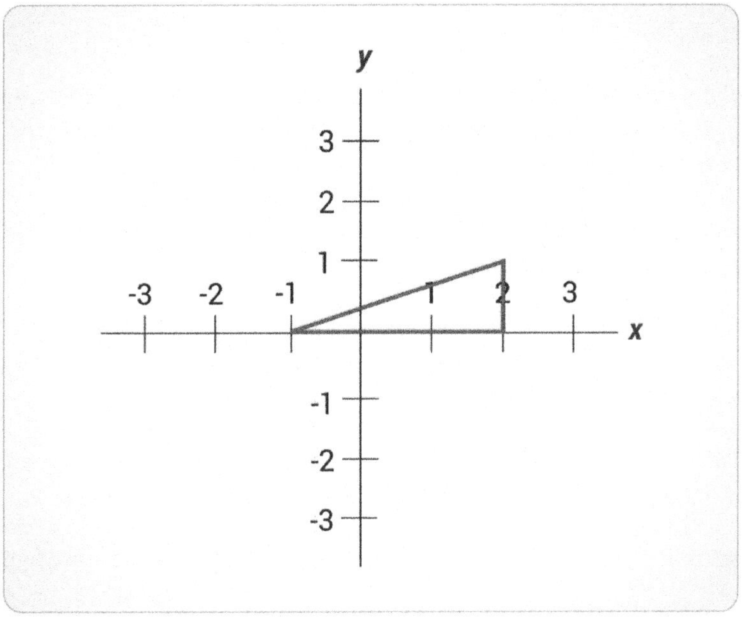

To do this, flip the *x* values of the points involved to the negatives of themselves, while keeping the *y* values the same. The image is shown here.

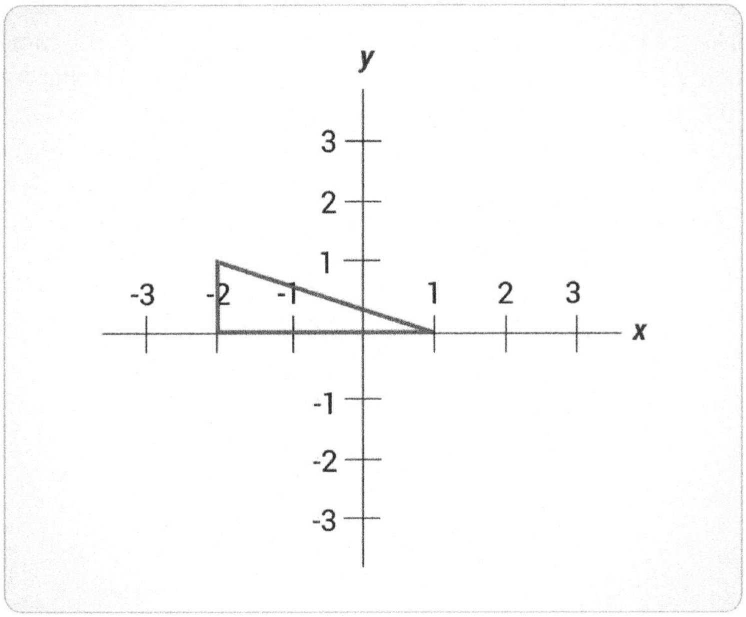

The new vertices will be (1, 0), (-2, 1), and (-2, 0).

For rotation, pick a center point, then rotate every vertex along a circle around that point by the same angle. This procedure is also not easy to express in Cartesian coordinates, and this is not a requirement on this test. However, as with reflections, it's helpful to draw the figures and see what the result of the rotation would look like. This transformation can be performed using a compass and protractor.

Each one of these transformations can be performed on the coordinate plane without changes to the original dimensions or angles.

If two figures in the plane involve the same distances and angles, they are called **congruent figures**. In other words, two figures are congruent when they go from one form to another through reflection, rotation, and translation, or a combination of these.

Remember that rotation and translation will give back a new figure that is identical to the original figure, but reflection will give back a mirror image of it.

To recognize that a figure has undergone a rotation, check to see that the figure has not been changed into a mirror image, but that its orientation has changed (that is, whether the parts of the figure now form different angles with the x and y axes).

To recognize that a figure has undergone a translation, check to see that the figure has not been changed into a mirror image, and that the orientation remains the same.

To recognize that a figure has undergone a reflection, check to see that the new figure is a mirror image of the old figure.

Keep in mind that sometimes a combination of translations, reflections, and rotations may be performed on a figure.

A shape is dilated, or a **dilation** occurs, when each side of the original image is multiplied by a given scale factor. If the scale factor is less than 1 and greater than 0, the dilation contracts the shape, and the resulting shape is smaller. If the scale factor equals 1, the resulting shape is the same size, and the dilation is a rigid motion. Finally, if the scale factor is greater than 1, the resulting shape is larger, and the dilation expands the shape. The **center of dilation** is the point where the distance from it to any point on the new shape equals the scale factor times the distance from the center to the corresponding point in the pre-image. Dilation isn't an isometric transformation because distance isn't preserved. However, angle measure, parallel lines, and points on a line all remain unchanged.

The following figure is an example of translation, rotation, dilation, and reflection:

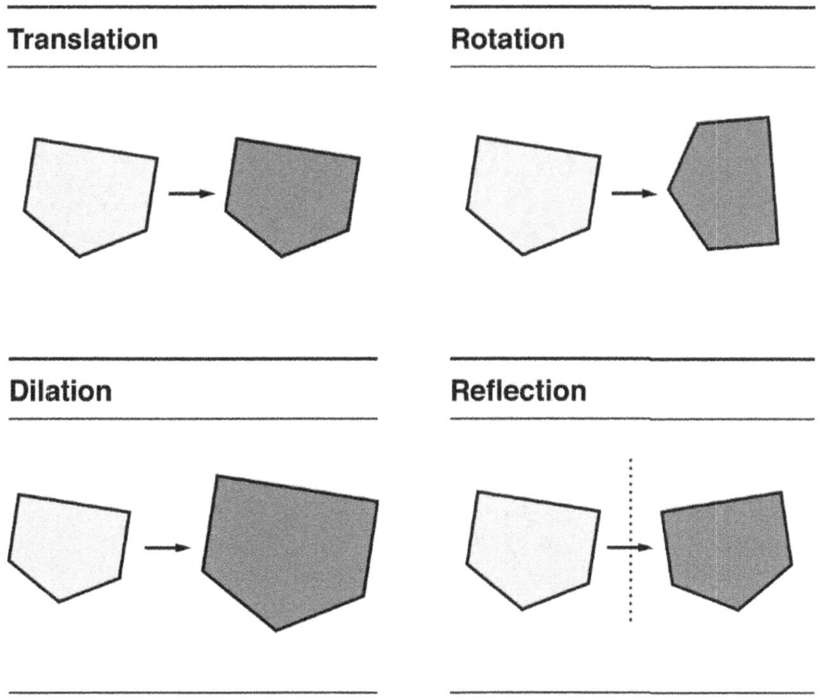

Differentiating Between Transformations That Preserve Congruence and Those That Do Not

Two figures are congruent if there is a rigid motion that can map one figure onto the other. Therefore, all pairs of sides and angles within the image and pre-image must be congruent. For example, in triangles, each pair of the three sides and three angles must be congruent. Similarly, in two four-sided figures, each pair of the four sides and four angles must be congruent.

Two figures are similar if there is a combination of translations, reflections, rotations, and dilations, which maps one figure onto the other. The difference between congruence and similarity is that dilation can be used in similarity. Therefore, side lengths between each shape can differ. However, angle measure must be preserved within this definition. If two polygons differ in size so that the lengths of corresponding line segments differ by the same factor, but corresponding angles have the same measurement, they are similar.

The Effect of Dilations, Translations, Reflections, and Rotations on Coordinates

As mentioned, a shape is dilated, or a **dilation** occurs, when each side of the original image is multiplied by a given scale factor. If the scale factor is less than 1 and greater than 0, the dilation contracts the shape, and the resulting shape is smaller. If the scale factor equals 1, the resulting shape is the same size, and the dilation is a rigid motion. Finally, if the scale factor is greater than 1, the resulting shape is larger, and the dilation expands the shape. The **center of dilation** is the point where the distance from it to any point on the new shape equals the scale factor times the distance from the center to the corresponding point in the pre-image. Dilation isn't an isometric transformation because distance isn't preserved. However, angle measure, parallel lines, and points on a line all remain unchanged.

The easiest example is to dilate around the origin. In this case, multiply the *x* and *y* coordinates by a **scale factor**, k, sending points (x, y) to (kx, ky).

As an example, draw a dilation of the following triangle, whose vertices will be the points (-1, 0), (1, 0), and (1, 1).

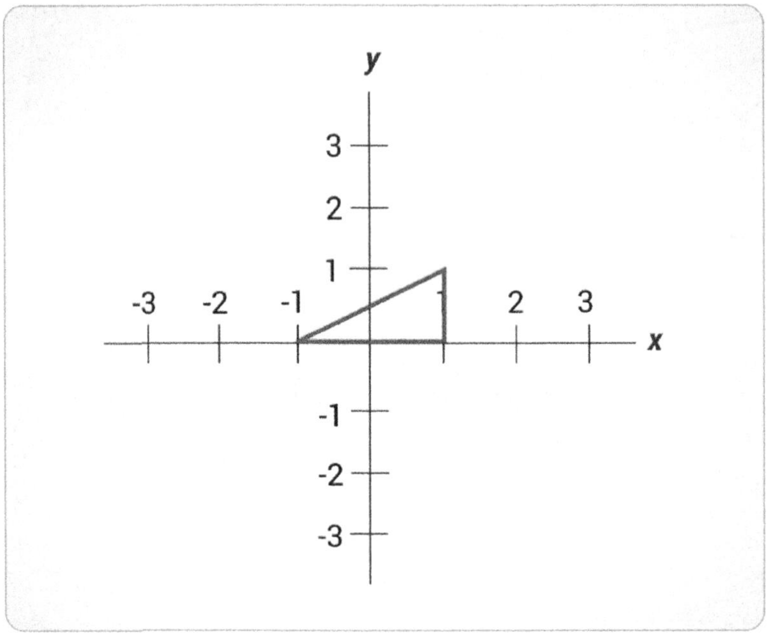

For this problem, dilate by a scale factor of 2, so the new vertices will be (-2, 0), (2, 0), and (2, 2).

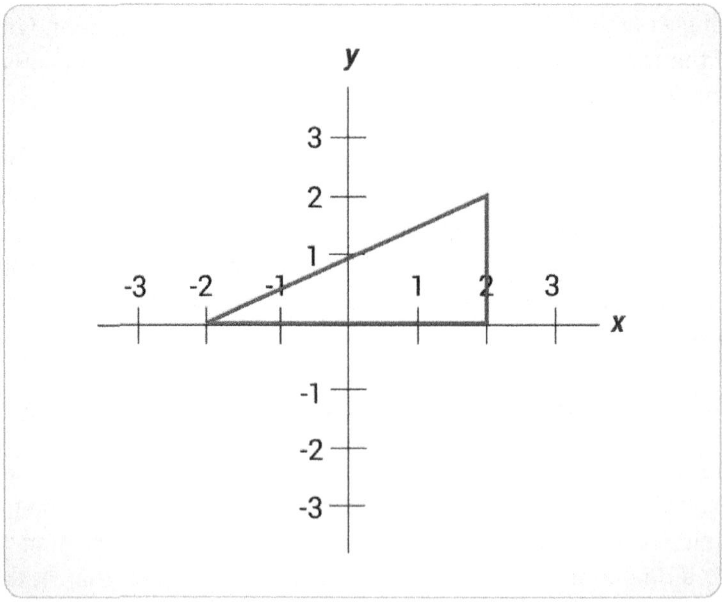

Let's look at another dilation in the coordinate plane. Consider a rectangle that has been plotted with four corners being the following coordinates: (-4,0), (3,0), (3,-3), and (-4,-3). Now, consider a dilation with a scale factor of 2, meaning all 4 sides will double their length. The resulting is an enlargement of the original image. Their new coordinates of the corners of the rectangle will be (-4,0), (6,6), (6,-6), and (-8,-6). Notice that within all four new corners, each coordinate has doubled. Also, the side lengths have doubled from 3 and 7 units to 6 and 14 units. This process can be seen in the following picture where the figure on the left represents the original image and the figure on the right represents the dilation:

Original

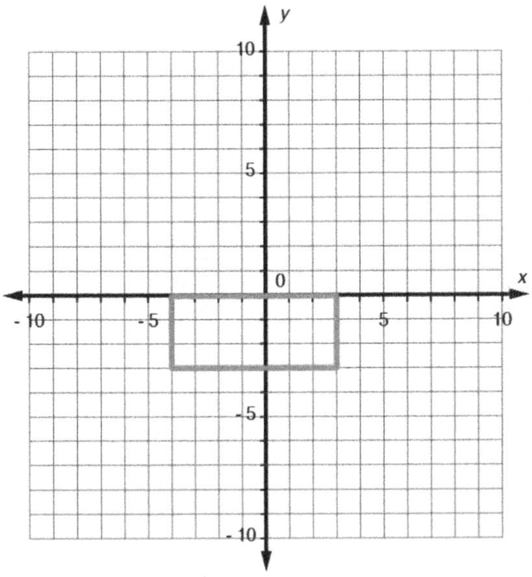

Dilation

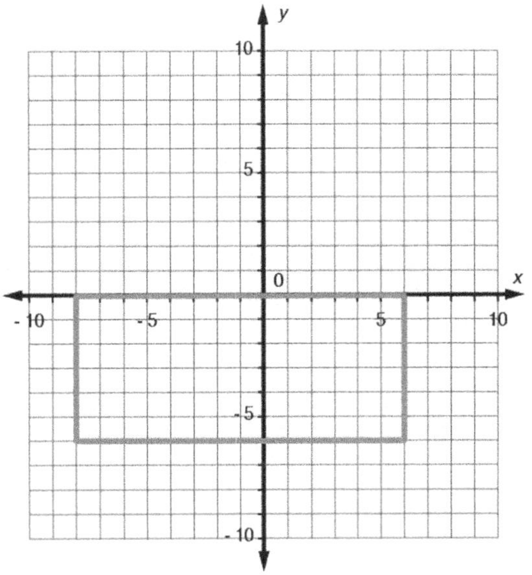

A translation is a transformation onto a shape that does not change its size and orientation. Therefore, consider a shape that is plotted on a coordinate plane. If it contains a point (x,y), then if the transformation shifts the figure to the right 6 units, you would need to add 6 to the x-coordinate. If the transformation shifts the figure to the left 5 units, you would need to subtract 5 from the x-coordinate. Similarly, if the transformation shifts the figure up 4 units, you would need to add 4 to the y-coordinate and if the transformation shifts the figure down 3 units, you would need to subtract 3 from the y-coordinate.

A reflection is a transformation onto a shape that reflects the figure and does not change its shape or size. In this case, its position and orientation changes. Consider any point (x,y) on a figure within a coordinate plane. If the figure is reflected over the x-axis, the new coordinate would be (x,-y). If the figure is reflected over the y-axis, the new coordinate would be (-x,y). This reflection changes what quadrant each point is in.

Finally, a rotation is a transformation onto a shape is when the position changes, but the orientation, size, and shape do not change. If a point (x,y) on a shape in a coordinate plane is rotated 90° counterclockwise about the origin, the new coordinates are (-y,x). Similarly, if the point is rotated 180° counterclockwise about the origin, the new coordinates are (-x,-y). Thirdly, if the point is rotated 270°

counterclockwise about the origin, the new coordinates are (y,-x). Finally, if the point is rotated 360° counterclockwise about the origin, the rotation just brings it back to its current position at (x,y).

Here is a picture of a 90° rotation about the origin:

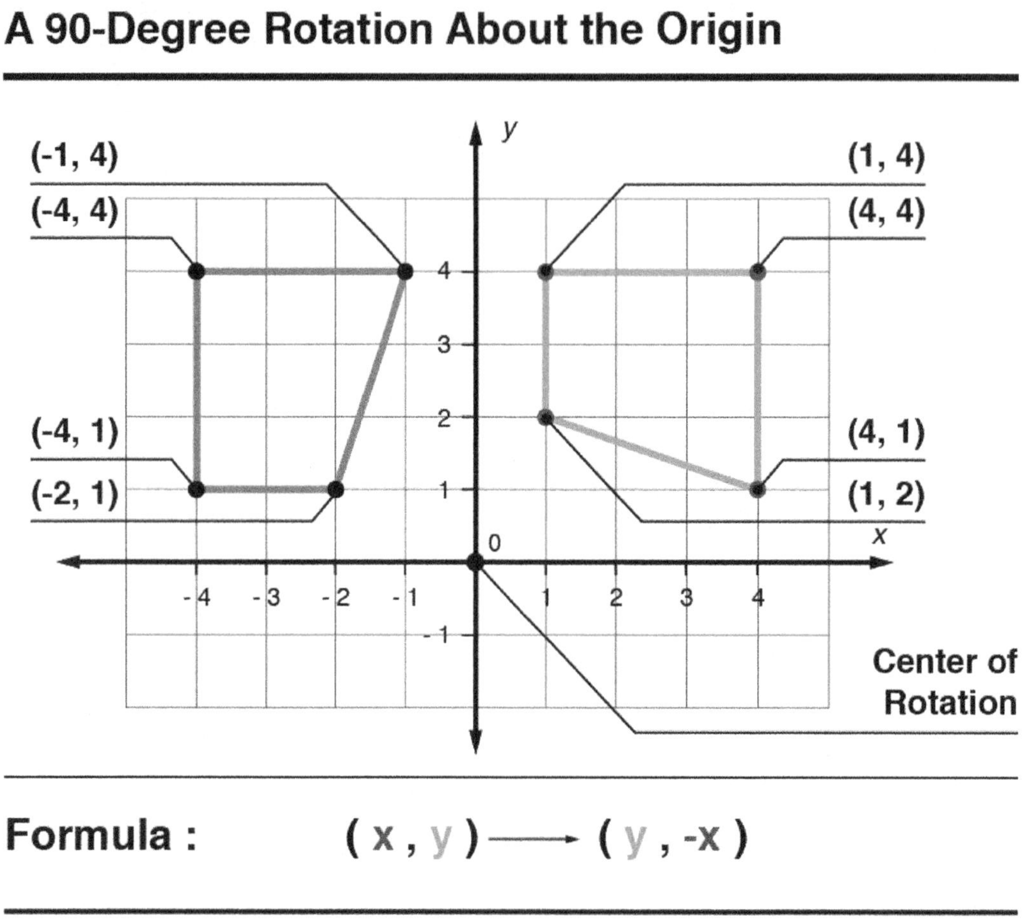

<u>Using Algebraic Representations to Explain the Effect of Scale Factors on Dilations</u>
Consider a shape on a coordinate plane in which any point on the shape can be labeled as (x,y) where x is the x-coordinate and y is the y-coordinate. If the origin is used as the center of dilation and the scale factor is positive, call it k, then a corresponding point on the graph of the dilation is (kx,ky). What that means is that each coordinate in the original figure is just multiplied times the scale factor, resulting in a corresponding coordinate in the new figure of the dilation.

This was the case in the rectangle shown above and is also the case in the octagon shown below:

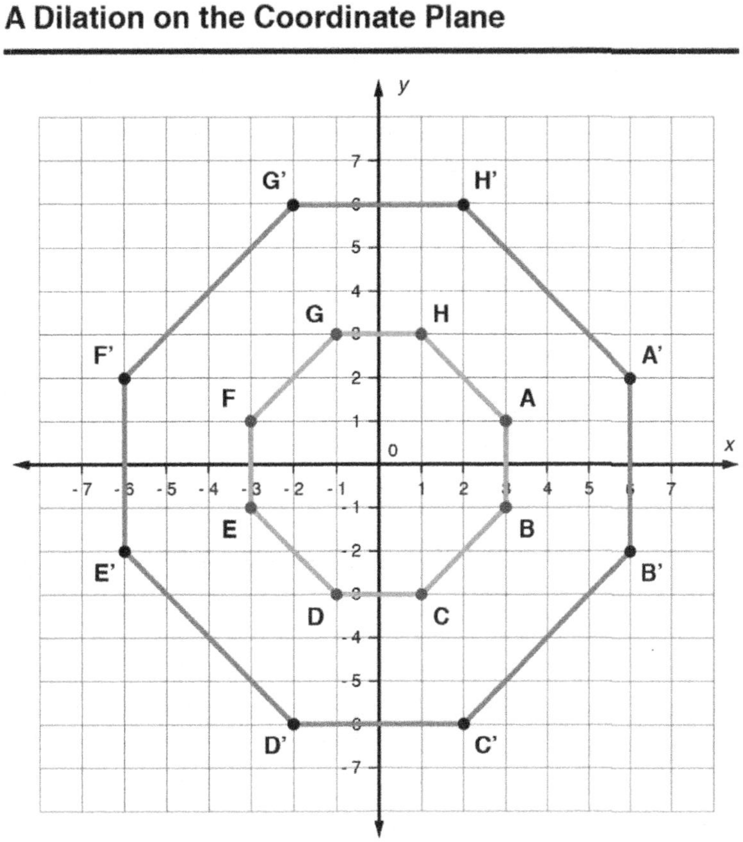

Consider the coordinate (3,1) which represents point A. Its corresponding point in the dilation is A' with coordinates (6,2). Each point has been multiplied times 2. Similarly, G, with ordered pair (-1,3), and G', with ordered pair (-2,6), differ by a multiplicative factor of 2 as well because -1 (the x-coordinate of the first ordered pair) times 2 is -2 (the x-coordinate of the first ordered pair) and 3 (the y-coordinate of the first ordered pair) times 2 is 6 (the y-coordinate of the first ordered pair).

Similar Figures

Note that after a dilation, the distances between the vertices of the figure will have changed, but the angles remain the same. The two figures that are obtained by dilation, along with possibly translation, rotation, and reflection, are all **similar** to one another. Another way to think of this is that similar figures have the same number of vertices and edges, and their angles are all the same. Similar figures have the same basic shape but are different in size.

Again, when a shape can be obtained from resizing another shape, meaning compressing (making it smaller) or expanding (making it bigger), the shapes can be labeled as similar. Another term for a resize is a dilation, in which the end result is the same shape as an original image but a different size. Every dilation includes a scale factor. If the scale factor is greater than 1, the result is an enlargement of the original shape. If the scale factor is less than 1, the result is a reduction of the original shape. If the scale factor is equal to 1, the result is the same shape. It is true that with any dilation, the ratio of corresponding sides of the two shapes are proportional.

Consider the following two hexagons:

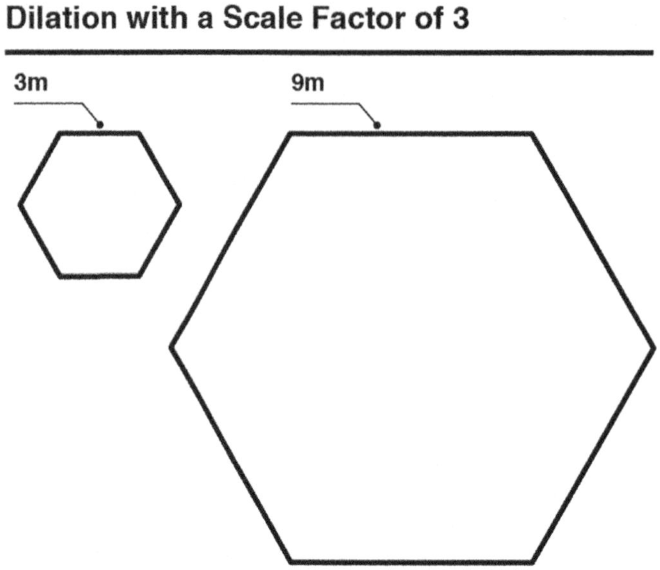

The second hexagon is 3 times larger than the first, so the scale factor of the dilation is 3. The first hexagon had each side length tripled, which resulted in an enlargement. Note that every side of the second hexagon is 3 times as long as the corresponding sides in the first hexagon; therefore, they are proportional.

With any two figures that are similar but have different dimensions, if the ratio of two corresponding sides is known, then that ratio, or scale factor, holds true for all of the dimensions of the new figure.

Here is an example of applying this principle. Suppose that Lara is 5 feet tall and is standing 30 feet from the base of a light pole, and her shadow is 6 feet long. How high is the light on the pole? To figure this, it helps to make a sketch of the situation:

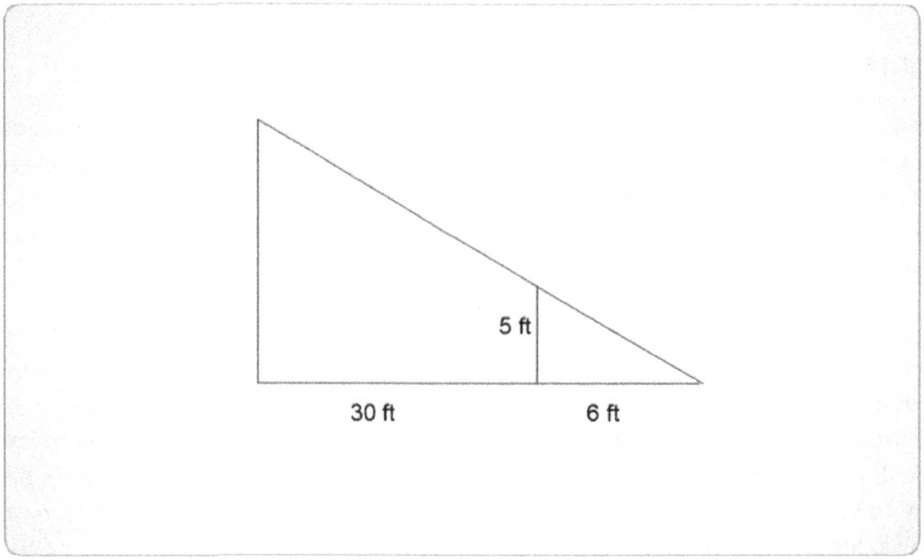

The light pole is the left side of the triangle. Lara is the 5-foot vertical line. Notice that there are two right triangles here, and that they have all the same angles as one another. Therefore, they form similar triangles. So, figure the ratio of proportionality between them.

The bases of these triangles are known. The small triangle, formed by Lara and her shadow, has a base of 6 feet. The large triangle, formed by the light pole along with the line from the base of the pole out to the end of Lara's shadow is $30 + 6 = 36$ feet long. So, the ratio of the big triangle to the little triangle will be $\frac{36}{6} = 6$. The height of the little triangle is 5 feet. Therefore, the height of the big triangle will be $6 \times 5 = 30$ feet, meaning that the light is 30 feet up the pole.

Notice that the perimeter of a figure changes by the ratio of proportionality between two similar figures, but the area changes by the *square* of the ratio. This is because if the length of one side is doubled, the area is quadrupled.

As an example, suppose two rectangles are similar, but the edges of the second rectangle are three times longer than the edges of the first rectangle. The area of the first rectangle is 10 square inches. How much more area does the second rectangle have than the first?

To answer this, note that the area of the second rectangle is $3^2 = 9$ times the area of the first rectangle, which is 10 square inches. Therefore, the area of the second rectangle is going to be $9 \times 10 = 90$ square inches. This means it has $90 - 10 = 80$ square inches more area than the first rectangle.

As a second example, suppose *X* and *Y* are similar right triangles. The hypotenuse of *X* is 4 inches. The area of *Y* is $\frac{1}{4}$ the area of *X*. What is the hypotenuse of *Y*?

First, realize the area has changed by a factor of $\frac{1}{4}$. The area changes by a factor that is the *square* of the ratio of changes in lengths, so the ratio of the lengths is the square root of the ratio of areas. That means that the ratio of lengths must be is $\sqrt{\frac{1}{4}} = \frac{1}{2}$, and the hypotenuse of *Y* must be $\frac{1}{2} \times 4 = 2$ inches.

Volumes between similar solids change like the cube of the change in the lengths of their edges. Likewise, if the ratio of the volumes between similar solids is known, the ratio between their lengths is known by finding the cube root of the ratio of their volumes.

For example, suppose there are two similar rectangular pyramids *X* and *Y*. The base of *X* is 1 inch by 2 inches, and the volume of *X* is 8 inches. The volume of *Y* is 64 inches. What are the dimensions of the base of *Y*?

To answer this, first find the ratio of the volume of *Y* to the volume of *X*. This will be given by $\frac{64}{8} = 8$. Now the ratio of lengths is the cube root of the ratio of volumes, or $\sqrt[3]{8} = 2$. So, the dimensions of the base of *Y* must be 2 inches by 4 inches.

Properties of Angles and Lines in Triangles

Parallel, Perpendicular, and Intersecting Lines

Two lines are parallel if they have the same slope and a different intercept. Two lines are **perpendicular** if the product of their slope equals -1. Parallel lines never intersect unless they are the same line, and perpendicular lines intersect at a right angle. If two lines aren't parallel, they must intersect at one point. If lines do cross, they're labeled as **intersecting lines** because they "intersect" at one point. If they intersect at more than one point, they're the same line. Determining equations of lines based on properties of parallel and perpendicular lines appears in word problems. To find an equation of a line, both the slope and a point the line goes through are necessary. Therefore, if an equation of a line is needed that's parallel to a given line and runs through a specified point, the slope of the given line and the point are plugged into the point-slope form of an equation of a line. Secondly, if an equation of a line is needed that's perpendicular to a given line running through a specified point, the negative reciprocal of the slope of the given line and the point are plugged into the **point-slope form**. Also, if the point of intersection of two lines is known, that point will be used to solve the set of equations. Therefore, to solve a system of equations, the point of intersection must be found. If a set of two equations with two unknown variables has no solution, the lines are parallel.

The **Parallel Postulate** states that if two parallel lines are cut by a transversal, then the corresponding angles are equal. Here is a picture that highlights this postulate:

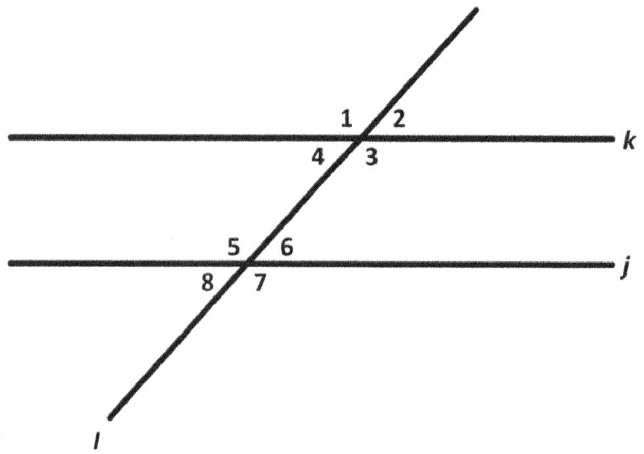

Because lines *k* and *i* are parallel, when cut by transversal *l*, angles 1 and 5 are equal, angles 2 and 6 are equal, angles 4 and 8 are equal, and angles 3 and 7 are equal. Note that angles 1 and 2, 3 and 4, 5 and 6, and 7 and 8 add up to 180 degrees.

This statement is equivalent to the **Alternate Interior Angle Theorem**, which states that when two parallel lines are cut by a transversal, the resultant interior angles are congruent. In the picture above, angles 3 and 5 are congruent, and angles 4 and 6 are congruent.

The Parallel Postulate or the Alternate Interior Angle Theorem can be used to find the missing angles in the following picture:

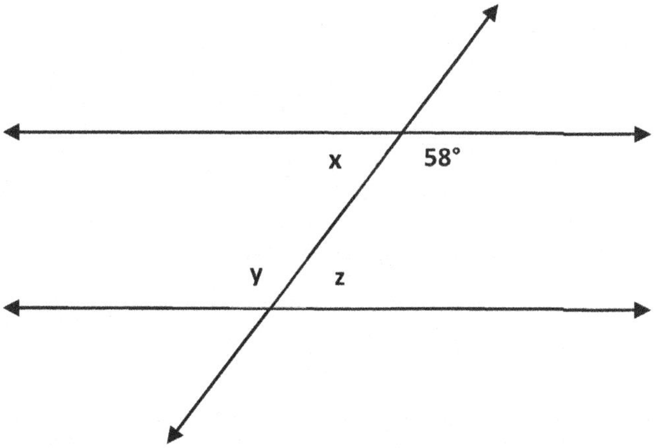

Assuming that the lines are parallel, angle x is found to be 122 degrees. Angle x and the 58-degree angle add up to 180 degrees. The Alternate Interior Angle Theorem states that angle y is equal to 58 degrees. Also, angles y and z add up to 180 degrees, so angle z is 122 degrees. Note that angles x and z are also alternate interior angles, so their equivalence can be used to find angle z as well.

An equivalent statement to the Parallel Postulate is that the sum of all angles in a triangle is 180 degrees. Therefore, given any triangle, if two angles are known, the third can be found accordingly.

Applying Angle Relationships to Solve Problems

An **angle** consists of two rays that have a common endpoint. This common endpoint is called the **vertex** of the angle. The two rays can be called sides of the angle. The angle below has a vertex at point *B* and the sides consist of ray *BA* and ray *BC*. An angle can be named in three ways:

1. Using the vertex and a point from each side, with the vertex letter in the middle.
2. Using only the vertex. This can only be used if it is the only angle with that vertex.
3. Using a number that is written inside the angle.

The angle below can be written $\angle ABC$ (read angle *ABC*), $\angle CBA$, $\angle B$, or $\angle 1$.

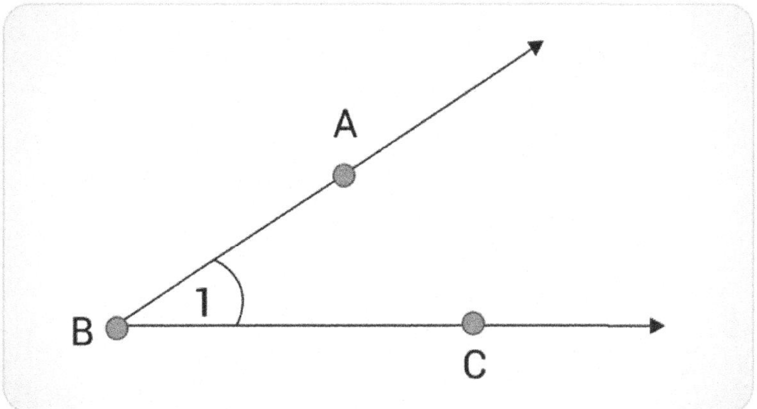

An angle divides a **plane**, or flat surface, into three parts: the angle itself, the interior (inside) of the angle, and the exterior (outside) of the angle. The figure below shows point *M* on the interior of the angle and point *N* on the exterior of the angle.

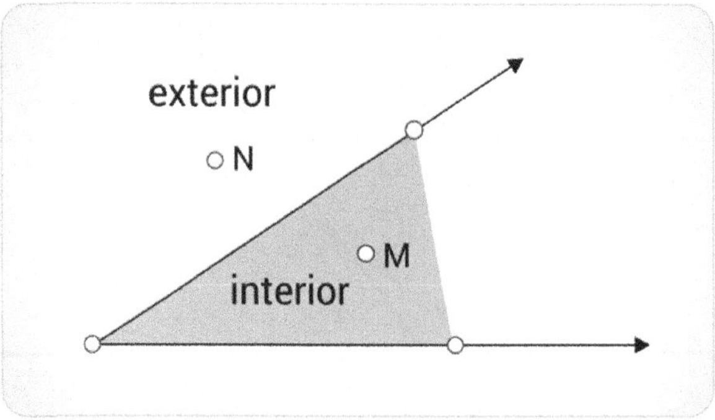

Angles can be measured in units called degrees, with the symbol °. The degree measure of an angle is between 0° and 180°, is a measure of rotation, and can be obtained by using a **protractor**, which is the tool pictured below.

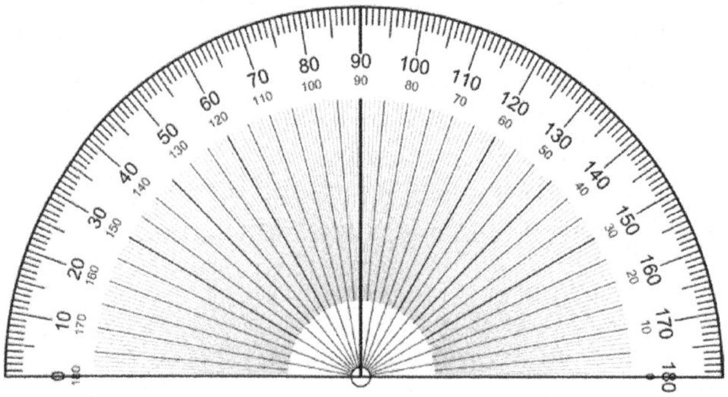

To use a protractor to measure an angle, the vertex, or corner, of the angle goes in the midpoint of the protractor, in that small circle along the bottom straight edge. Then one line of the angle is lined up along the bottom edge, toward the 0° indicator. Then, the degrees are read wherever the other line of the angle crosses.

In the example below, it can be seen that the angle measures about 30°:

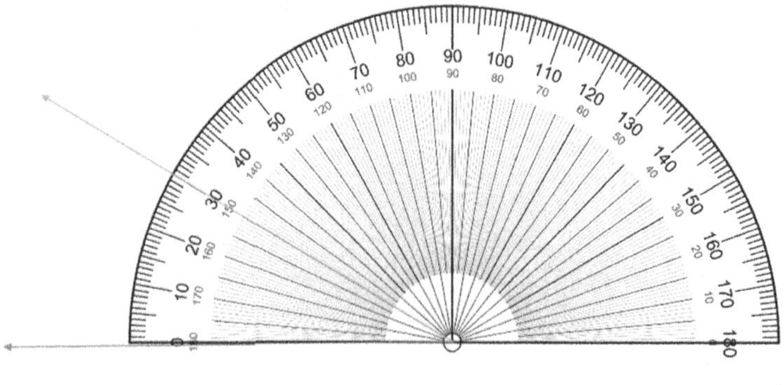

A straight angle (or simply a line) measures exactly 180°. A right angle's sides meet at the vertex to create a square corner. A right angle measures exactly 90° and is typically indicated by a box drawn in the interior of the angle. An acute angle has an interior that is narrower than a right angle. The measure of an acute angle is any value less than 90° and greater than 0°. For example, 89.9°, 47°, 12°, and 1°. An obtuse angle has an interior that is wider than a right angle. The measure of an obtuse angle is any value greater than 90° but less than 180°. For example, 90.1°, 110°, 150°, and 179.9°. Any two angles that sum up to 90 degrees are known as **complementary angles**.

- Acute angles: Less than 90°
- Obtuse angles: Greater than 90°
- Right angles: 90°
- Straight angles: 180°

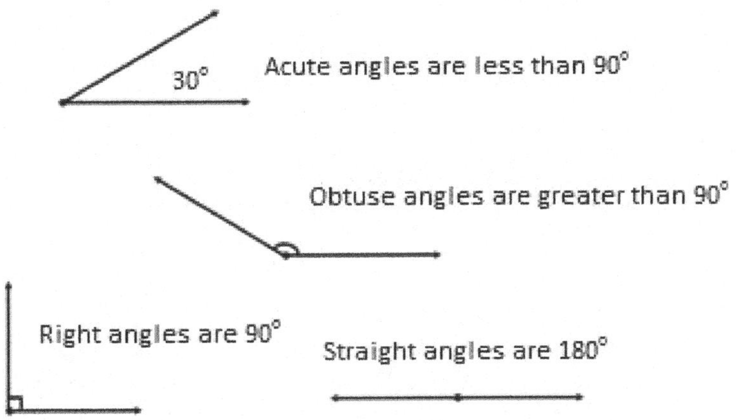

Supplementary angles add up to 180 degrees. **Vertical angles** are two nonadjacent angles formed by two intersecting lines. For example, in the following picture, angles 4 and 2 are vertical angles and so are angles 1 and 3:

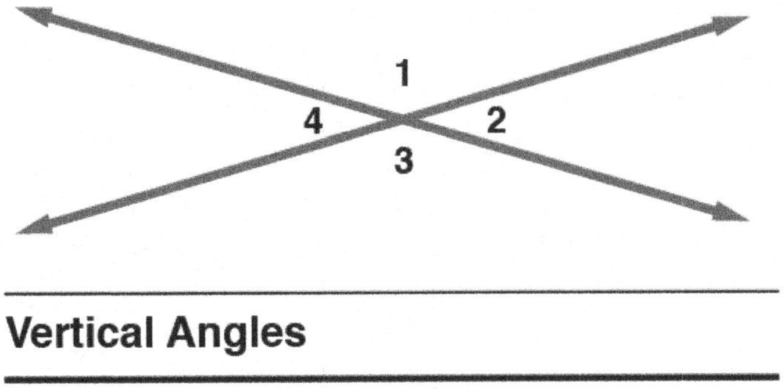

Corresponding angles are two angles in the same position whenever a straight line (known as a **transversal**) crosses two others. If the two lines are parallel, the corresponding angles are equal. In the following diagram, angles 1 and 3 are corresponding angles but aren't equal to each other:

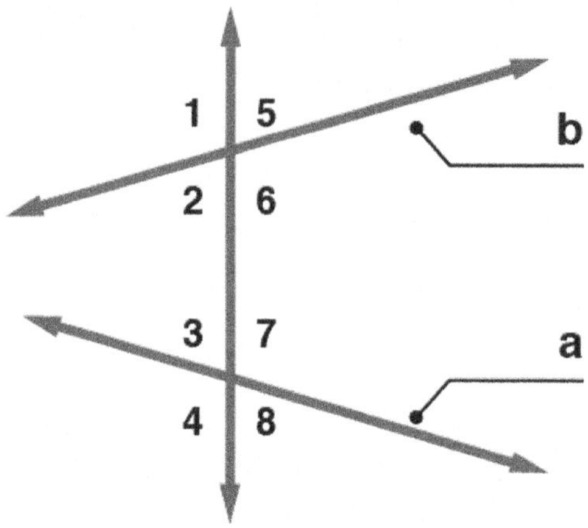

Corresponding Angles

Alternate interior angles are also a pair of angles formed when two lines are crossed by a transversal. They are opposite angles that exist inside of the two lines. In the corresponding angles diagram above, angles 2 and 7 are alternate interior angles, as well as angles 6 and 3. **Alternate exterior angles** are opposite angles formed by a transversal but, in contrast to interior angles, exterior angles exist outside the two original lines. Therefore, angles 1 and 8 are alternate exterior angles and so are angles 5 and 4. Finally, **consecutive interior angles** are pairs of angles formed by a transversal. These angles are located on the same side of the transversal and inside the two original lines. Therefore, angles 2 and 3 are a pair of consecutive interior angles, and so are angles 6 and 7. These definitions are instrumental in solving many problems that involve determining relationships between angles.

To determine angle measures for adjacent angles, angles that share a common side and vertex, at least one of the angles must be known. Other information that is necessary to determine such measures include that there are 90^{0} in a right angle, and there are 180^{0} in a straight line. Therefore, if two adjacent angles form a right angle, they will add up to 90^{0}, and if two adjacent angles form a straight line, they add up to 180^{0}.

If the measurement of one of the adjacent angles is known, the other can be found by subtracting the known angle from the total number of degrees.

For example, given the following situation, if angle *a* measures 55^0, find the measure of unknown angle *b*:

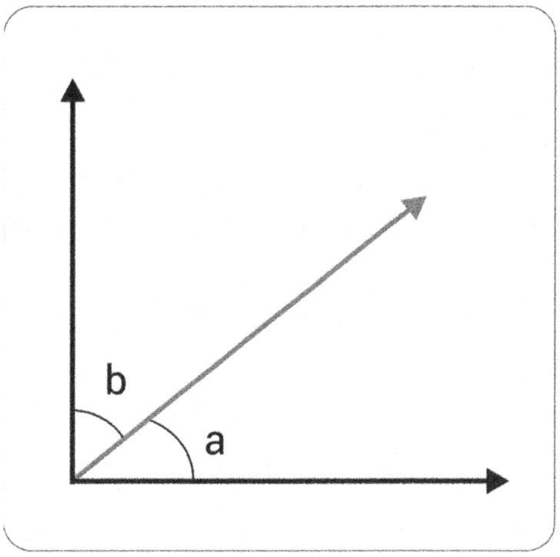

To solve this simply subtract the known angle measure from 90^0.

$$90° - 55° = 35°$$

The measure of *b* = 35°.

Given the following situation, if angle 1 measures 45^0, find the measure of the unknown angle 2:

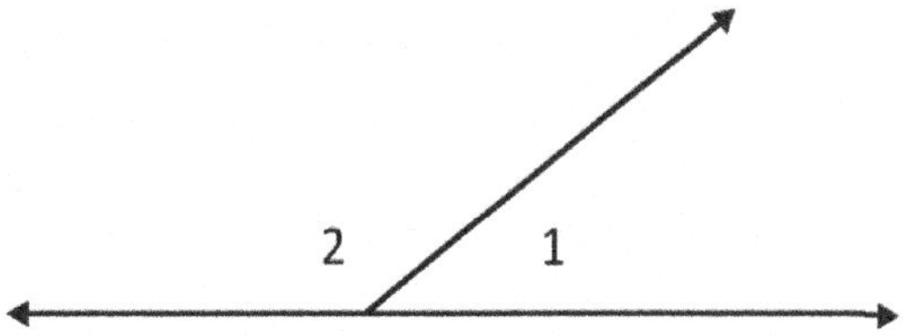

To solve this, simply subtract the known angle measure from 180^0.

$$180° - 45° = 135°$$

The measure of angle 2 = 135°.

In the case that more than two angles are given, use the same method of subtracting the known angles from the total measure.

For example, given the following situation, if angle *y* = 40^0, and angle *z* = 25^0, find unknown angle *x*.

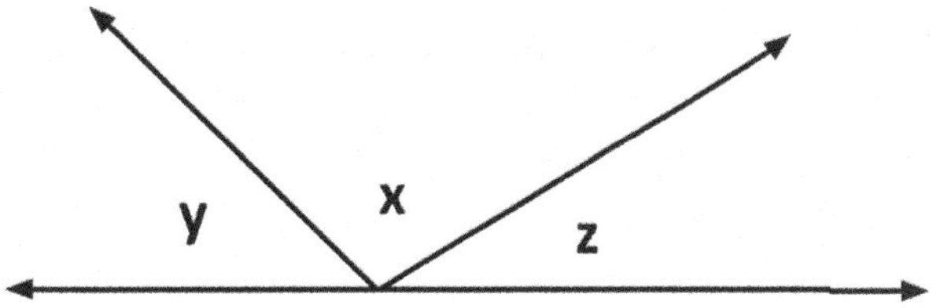

Subtract the known angles from 180°.

$$180° - 40° - 25° = 115°$$

The measure of angle $x = 115°$.

Relationships Between Sides and Angles in Triangles

The **Exterior Angle Theorem** states that the measure of an exterior angle of a triangle is equal to the sum of the measures of the two non-adjacent interior angles inside the triangle. It can be shown here:

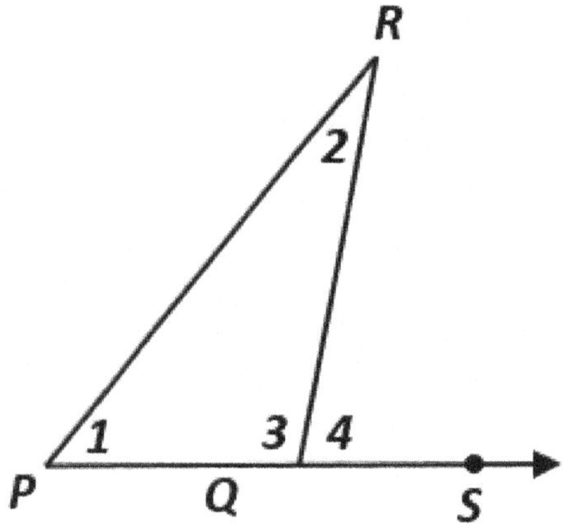

The measure of angle 4 is equal to the sum of the measures of angles 1 and 2. Also, the Triangle Inequality Theorem states that for any triangle, the sum of the lengths of two sides is greater than or equal to the length of the third side. For example, in the picture above, $PQ + QR \geq PR$.

Within similar triangles, corresponding sides are proportional, and angles are congruent. In addition, within similar triangles, the ratio of the side lengths is the same. This property is true even if side lengths are different. Within right triangles, trigonometric ratios can be defined for the acute angle within the triangle. The functions are defined through ratios in a right triangle. **Sine** of acute angle, A, is opposite over hypotenuse, **cosine** is adjacent over hypotenuse, and **tangent** is opposite over adjacent. Note that expanding or shrinking the triangle won't change the ratios. However, changing the angle measurements will alter the calculations.

Within a right triangle, two complementary angles exist because the third angle is always 90 degrees. In this scenario, the *sine* of one of the complementary angles is equal to the cosine of the other angle. The opposite is also true. This relationship exists because sine and cosine will be calculated as the ratios of the same side lengths.

Triangle Congruence

The criteria needed to prove triangles are congruent involves both angle and side congruence. Both pairs of related angles and sides need to be of the same measurement to use congruence in a proof. The criteria to prove similarity in triangles involves proportionality of side lengths. Angles must be congruent in similar triangles; however, corresponding side lengths only need to be a constant multiple of each other. Once similarity is established, it can be used in proofs as well. Relationships in geometric figures other than triangles can be proven using triangle congruence and similarity. If a similar or congruent triangle can be found within another type of geometric figure, their criteria can be used to prove a relationship about a given formula. For instance, a rectangle can be broken up into two congruent triangles.

If two angles of one triangle are congruent with two angles of a second triangle, the triangles are similar. This is because, within any triangle, the sum of the angle measurements is 180 degrees. Therefore, if two are congruent, the third angle must also be congruent because their measurements are equal. Three congruent pairs of angles mean that the triangles are similar.

There are five theorems to show that triangles are congruent when it's unknown whether each pair of angles and sides are congruent. Each theorem is a shortcut that involves different combinations of sides and angles that must be true for the two triangles to be congruent. For example, **side-side-side (SSS)** states that if all sides are equal, the triangles are congruent. **Side-angle-side (SAS)** states that if two pairs of sides are equal and the included angles are congruent, then the triangles are congruent.

Similarly, **angle-side-angle (ASA)** states that if two pairs of angles are congruent and the included side lengths are equal, the triangles are similar. **Angle-angle-side (AAS***)* states that two triangles are congruent if they have two pairs of congruent angles and a pair of corresponding equal side lengths that aren't included. Finally, **hypotenuse-leg (HL)** states that if two right triangles have equal hypotenuses and an equal pair of shorter sides, then the triangles are congruent. An important item to note is that angle-angle-angle *(AAA)* is not enough information to have congruence. It's important to understand why these rules work by using rigid motions to show congruence between the triangles with the given properties. For example, three reflections are needed to show why *SAS* follows from the definition of congruence.

Medians, Midpoints, and Altitudes

A **median** of a triangle is the line drawn from a vertex to the midpoint on the opposite side. A triangle has three medians, and their point of intersection is known as the **centroid**. An **altitude** is a line drawn from a vertex perpendicular to the opposite side. A triangle has three altitudes, and their point of intersection is known as the **orthocenter**. An altitude can actually exist outside, inside, or on the triangle depending on the placement of the vertex. Many problems involve these definitions. For example, given one endpoint of a line segment and the midpoint, the other endpoint can be determined by using the midpoint formula. In addition, area problems heavily depend on these definitions. For example, it can be proven that the median of a triangle divides it into two regions of equal areas. The actual formula for the area of a triangle depends on its altitude.

Special Triangles

An **isosceles triangle** contains at least two equal sides. Therefore, it must also contain two equal angles and, subsequently, contain two medians of the same length. An isosceles triangle can also be labelled as an **equilateral triangle** (which contains three equal sides and three equal angles) when it meets these conditions. In an equilateral triangle, the measure of each angle is always 60 degrees. Also, within an equilateral triangle, the medians are of the same length. A **scalene triangle** can never be an equilateral or an isosceles triangle because it contains no equal sides and no equal angles. Also, medians in a scalene triangle can't have the same length. However, a **right triangle**, which is a triangle containing a 90-degree angle, can be a scalene triangle.

There are two types of special right triangles. The **30-60-90 right triangle** has angle measurements of 30 degrees, 60 degrees, and 90 degrees. Because of the nature of this triangle, and through the use of the Pythagorean theorem, the side lengths have a special relationship. If x is the length opposite the 30-degree angle, the length opposite the 60-degree angle is $\sqrt{3}x$, and the hypotenuse has length $2x$. The

45-45-90 right triangle is also special as it contains two angle measurements of 45 degrees. It can be proven that, if x is the length of the two equal sides, the hypotenuse is $x\sqrt{2}$. The properties of all of these special triangles are extremely useful in determining both side lengths and angle measurements in problems where some of these quantities are given and some are not.

Applying the Pythagorean Theorem

A Proof of the Pythagorean Theorem and Its Converse

The **Pythagorean Theorem** can be used to find the length of a missing side on a right triangle. It states that for every right triangle, the square of the length of the hypotenuse is equal to the sum of the squares of the lengths of the remaining two sides. The hypotenuse is the longest side of a right triangle and is also the side opposite the right angle.

According to the diagram, $a^2 + b^2 = c^2$, where c represents the hypotenuse, and a and b represent the lengths of the remaining two sides of the right triangle.

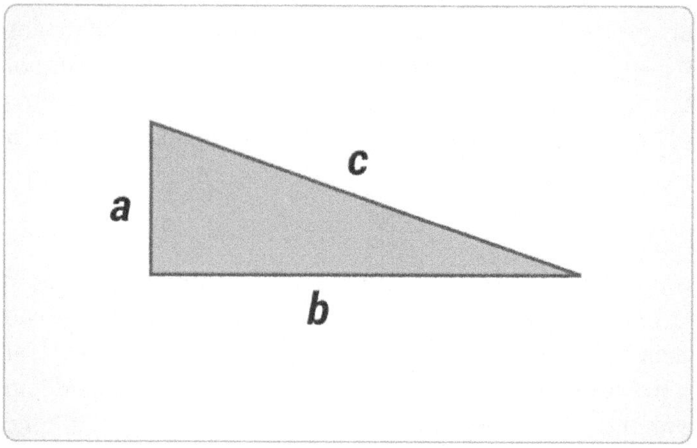

The **converse of the Pythagorean Theorem** is another important property of triangles, and it states that if the square of one side of a triangle is equal to the sum of the squares of the other two sides, then the triangle must have a 90-degree angle and therefore is a right triangle.

Trigonometric functions can be defined through ratios in a right triangle. **SOHCAHTOA** is a common acronym used to remember these ratios, which are defined by the relationships of the sides and angles relative to the right angle. Sine is opposite over hypotenuse, cosine is adjacent over hypotenuse, and tangent is opposite over adjacent. These ratios are the reciprocals of secant, cosecant, and cotangent, respectively. Angles can be measured in degrees or radians. Here is a diagram of SOHCAHTOA:

SOH $\sin\theta = \frac{\text{opposite}}{\text{hypotenuse}}$

CAH $\cos\theta = \frac{\text{adjacent}}{\text{hypotenuse}}$

TOA $\tan\theta = \frac{\text{opposite}}{\text{adjacent}}$

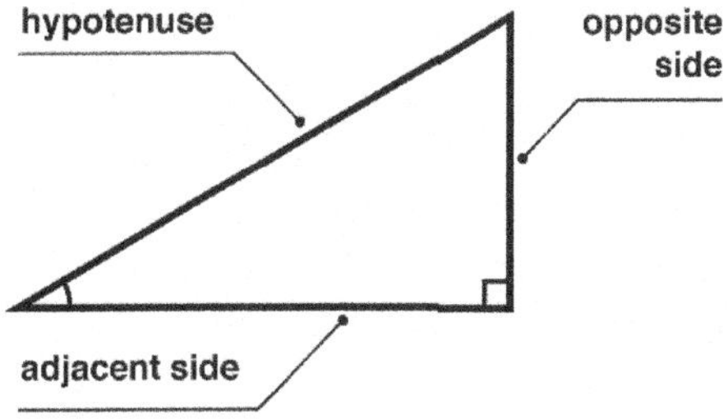

Applying the Pythagorean Theorem to Determine Unknown Side Lengths in Right Triangles

Again, the **Pythagorean theorem** is an important result in geometry. It states that for right triangles, the sum of the squares of the two shorter sides will be equal to the square of the longest side (also called the **hypotenuse**). The longest side will always be the side opposite to the 90° angle. If this side is called c, and the other two sides are a and b, then the Pythagorean theorem states that $c^2 = a^2 + b^2$. Since lengths are always positive, this also can be written as $c = \sqrt{a^2 + b^2}$.

The Pythagorean Theorem may be applied a multitude of ways. For example, a person wishes to build a garden in the shape of a rectangle, having the dimensions of 5 feet by 8 feet. The garden's design includes a diagonal board to separate various types of plants. The Pythagorean Theorem can be used to determine the length of the diagonal board.

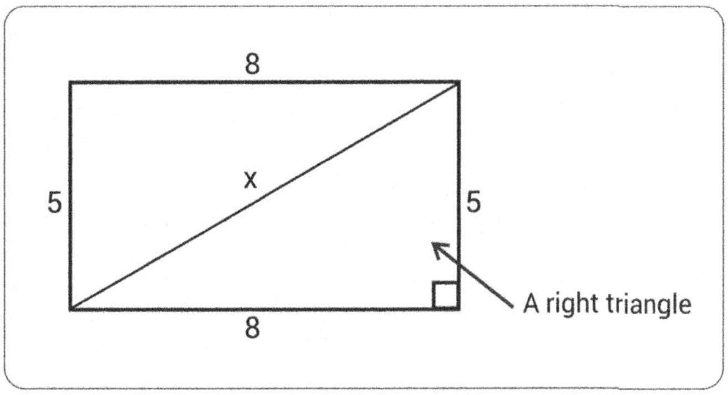

$a^2 + b^2 = c^2$

$5^2 + 8^2 = c^2$

$25 + 64 = c^2$

$c = \sqrt{89}$

$c = 9.43$

As another example of using the theorem, suppose that Shirley has a rectangular field that is 5 feet wide and 12 feet long, and she wants to split it in half using a fence that goes from one corner to the opposite corner. How long will this fence need to be? To figure this out, note that this makes the field into two right triangles, whose hypotenuse will be the fence dividing it in half. Therefore, the fence length will be given by:

$$\sqrt{5^2 + 12^2} = \sqrt{169} = 13 \text{ feet long}$$

Both the trigonometric functions and the Pythagorean theorem can be used in problems that involve finding either a missing side or a missing angle of a right triangle. To do so, one must look to see what sides and angles are given and select the correct relationship that will help find the missing value. These relationships can also be used to solve application problems involving right triangles. Often, it's helpful to draw a figure to represent the problem to see what's missing.

Applying the Pythagorean Theorem to Find the Distance Between Two Points in a Coordinate System

The distance between two points or the side length of a shape plotted in the coordinate plane can be found by plugging the endpoints into the distance formula:

$$d = \sqrt{(x_2 - x_1)^2 + (y_2 - y_1)^2}$$

between two ordered pairs (x_1, y_1) and (x_2, y_2). The distance formula is derived from the Pythagorean theorem. Once the side lengths are found, they can be added together to obtain the perimeter of the given polygon. Simplifications can be made for specific shapes such as squares and equilateral triangles. For example, one side length can be multiplied by 4 to obtain the perimeter of a square. Also, one side length can be multiplied by 3 to obtain the perimeter of an equilateral triangle. A similar technique can be used to calculate areas. For polygons, both side length and height can be found by using the same distance formula.

A **circle** can be defined as the set of all points that are the same distance (known as the radius, *r*) from a single point *C* (known as the center of the circle). The center has coordinates (h, k), and any point on the circle can be labelled with coordinates (x, y). As shown below, a **right triangle** is formed with these two points:

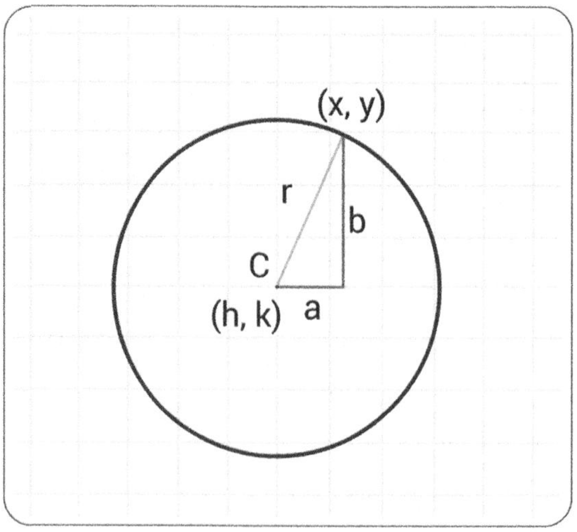

The Pythagorean theorem states that $a^2 + b^2 = r^2$. However, *a* can be replaced by $|x - h|$ and *b* can be replaced by $|y - k|$ by using the **distance formula**. That substitution results in $(x - h)^2 + (y - k)^2 = r^2$, which is the formula for finding the equation of any circle with a center (h, k) and a radius *r*. Note that sometimes *c* is used instead of *r*.

Solving Volume Problems with Cylinders, Cones, and Spheres

Using the Volume Formulas of Cones, Cylinders, and Spheres to Solve Problems

Volume is the measurement of how much space an object occupies, like how much space is in the cube. Volume is useful in determining the space within a certain three-dimensional object. Volume can be calculated for a cube, rectangular prism, cylinder, pyramid, cone, and sphere. By knowing specific dimensions of the objects, the volume of the object is computed with these figures. The units for the volumes of solids can include cubic centimeters, cubic meters, cubic inches, and cubic feet.

Cones

The formula to calculate the volume of a circular cone is similar to the formula for the volume of a pyramid. The primary difference in determining the area of a cone is that a circle serves as the base of a cone. Therefore, the area of a circle is used for the cone's base.

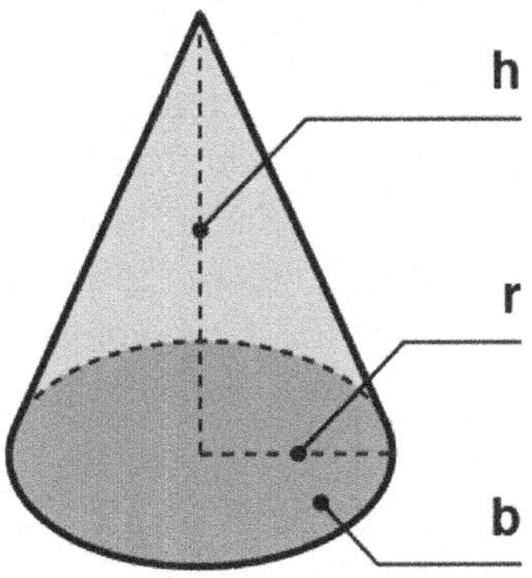

$$V = \frac{1}{3}\pi r^2 h$$

The variable *r* represents the radius, and the variable *h* represents the height of the cone. The formula used to calculate the volume of a cone is $\frac{1}{3}\pi r^2 h$. Essentially, the area of the base of the cone is multiplied by the cone's height. In a real-life example where the radius of a cone is 2 meters and the height of a cone is 5 meters, the volume of the cone is calculated by utilizing the formula $\frac{1}{3}\pi 2^2 \times 5 = 21$. After substituting 3.14 for π, the volume is 785.4 ft^3.

For completeness, to calculate the volume of a pyramid, the area of the base of the pyramid is multiplied by the pyramid's height by $\frac{1}{3}$. The area of the base of the pyramid is found by multiplying the base length by the base width.

Therefore, the formula to calculate a pyramid's volume is $(L \times W \times H) \div 3$.

Cylinders

Discovering a cylinder's volume requires the measurement of the cylinder's base, length of the radius, and height. The height of the cylinder can be represented with variable *h*, and the radius can be represented with variable *r*.

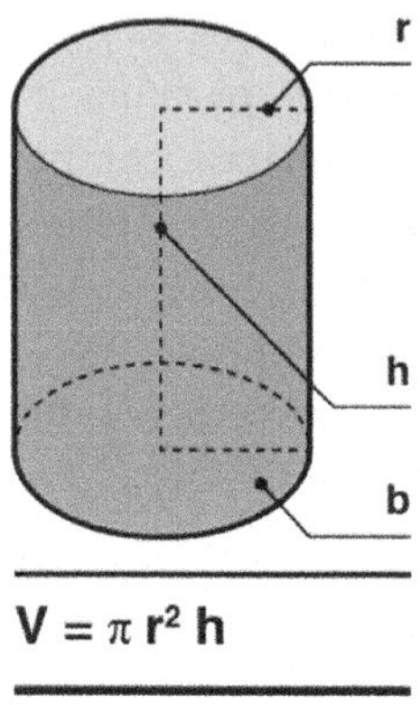

The formula to find the volume of a cylinder is $\pi r^2 h$. Notice that πr^2 is the formula for the area of a circle. This is because the base of the cylinder is a circle. To calculate the volume of a cylinder, the slices of circles needed to build the entire height of the cylinder are added together. For example, if the radius is 5 feet and the height of the cylinder is 10 feet, the cylinder's volume is calculated by using the following equation: $\pi 5^2 \times 10$. Substituting 3.14 for π, the volume is 785.4 ft^3.

Spheres

The volume of a sphere uses π due to its circular shape.

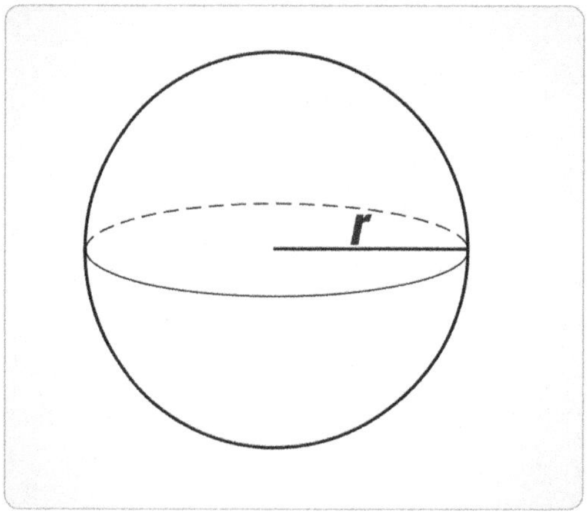

The length of the radius, *r*, is the only variable needed to determine the sphere's volume. The formula to calculate the volume of a sphere is $\frac{4}{3}\pi r^3$. Therefore, if the radius of a sphere is 8 centimeters, the volume of the sphere is calculated by utilizing the formula $\frac{4}{3}\pi(8)^3$ = 2,143 cm^3.

Problems involving volume, length, and other units can also be solved using ratios. If the following graphic of a cone is given, the problem may ask for the volume to be found.

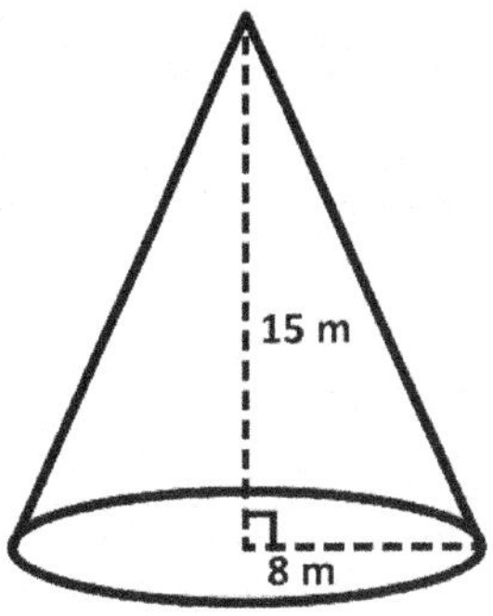

Referring to the formulas provided on the test, the volume of a cone is given as: $V = \pi r^2 \frac{h}{3}$, where *r* is the radius, and *h* is the height. Plugging $r = 8$ and $h = 15$ from the graphic into the formula, the following is obtained:

$$V = \pi(8^2)\frac{15}{3}$$

Therefore, volume of the cone is found to be 1005.3m^3. Sometimes, answers in different units are sought. If this problem wanted the answer in liters, $1005.3\ \text{m}^3$ would need to be converted. Using the equivalence statement $1\text{m}^3 = 1000\text{L}$, the following ratio would be used to solve for liters:

$$1005.3\text{m}^3 \times \frac{1000L}{1m^3}$$

Cubic meters in the numerator and denominator cancel each other out, and the answer is converted to 1,005,300 liters, or 1.0053×10^6 L.

Statistics and Probability

Patterns of Association in Bivariate Data

Constructing a Scatterplot and Using It to Describe Relationships Between Bivariate Data

A **scatterplot** displays the relationship between two variables. Values for the independent variable, typically denoted by *x*, are paired with values for the dependent variable, typically denoted by *y*. Each set of corresponding values are written as an ordered pair (*x*, *y*). To construct the graph, a coordinate grid is labeled with the *x*-axis representing the independent variable and the *y*-axis representing the dependent variable. Each ordered pair is graphed.

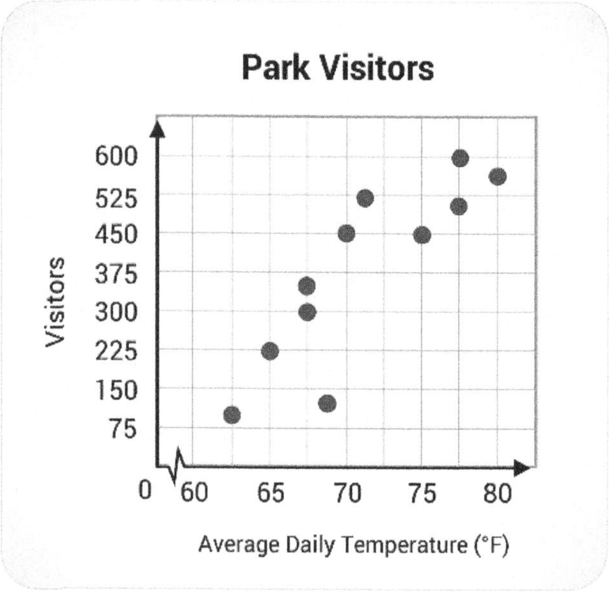

An **X-Y diagram**, also known as a **scatter diagram**, visually displays the relationship between two variables. The independent variable is placed on the **x-axis**, or horizontal axis, and the dependent variable is placed on the **y-axis**, or vertical axis.

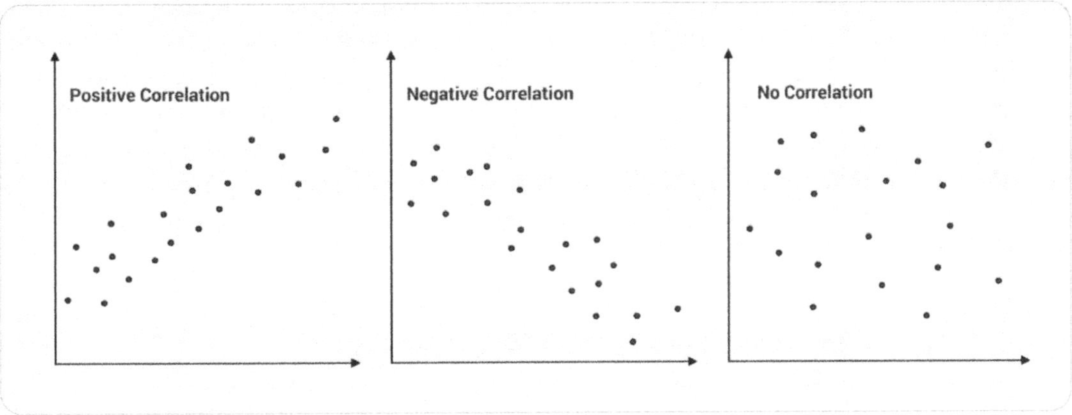

As shown in the figures above, an X-Y diagram may result in positive, negative, or no correlation between the two variables. In the first scatterplot, as the Y factor increases the X factor increases as well. The opposite is true as well: as the X factor increases the Y factor also increases. Thus, there is a positive correlation because one factor appears to positively affect the other factor

It's important to note, however, that a positive correlation between two variables doesn't equate to a cause-and-effect relationship. For example, a positive correlation between labor hours and units produced may not equate to a cause and effect relationship between the two. Any instance of correlation only indicates how likely the presence of one variable is in the instance of another. The variables should be further analyzed to determine which, if any, other variables (i.e. quality of employee work) may contribute to the positive correlation.

Lines of Best Fit in Linear Relationships

As mentioned, a **scatterplot** is a mathematical diagram that visually displays the relationship or connection between two variables. The independent variable is placed on the *x*-axis, or horizontal axis, and the dependent variable is placed on the *y*-axis, or vertical axis. When visually examining the points on the graph, if the points model a linear relationship, or a **line of best-fit** can be drawn through the points with the points relatively close on either side, then a correlation exists. If the line of best-fit has a positive slope (rises from left to right), then the variables have a positive correlation. If the line of best-fit has a negative slope (falls from left to right), then the variables have a negative correlation. If a line of best-fit cannot be drawn, then no correlation exists. A positive or negative correlation can be categorized as strong or weak, depending on how closely the points are graphed around the line of best-fit.

Regression lines are a way to calculate a relationship between the independent variable and the dependent variable. A straight line means that there's a linear trend in the data. Technology can be used to find the equation of this line (e.g., a graphing calculator or Microsoft Excel®). In either case, all of the data points are entered, and a line is "fit" that best represents the shape of the data. Other functions used to model data sets include quadratic and exponential models.

The formula for a linear regression line of best fit is $y = mx + b$, where m is the slope and b is the *y*-intercept. Both the slope and *y*-intercept are found in the **Method of Least Squares**, which is the process of finding the equation of the line through minimizing residuals. The slope represents the rate of change in *y* as x gets larger. Therefore, because *y* is the dependent variable, the slope actually provides the predicted values given the independent variable. The *y*-intercept is the predicted value for when the independent variable equals zero. In the temperature example, the *y*-intercept is the expected number of beach visitors for a very cold average daily temperature of zero degrees.

Here is an example of a data set and its regression line:

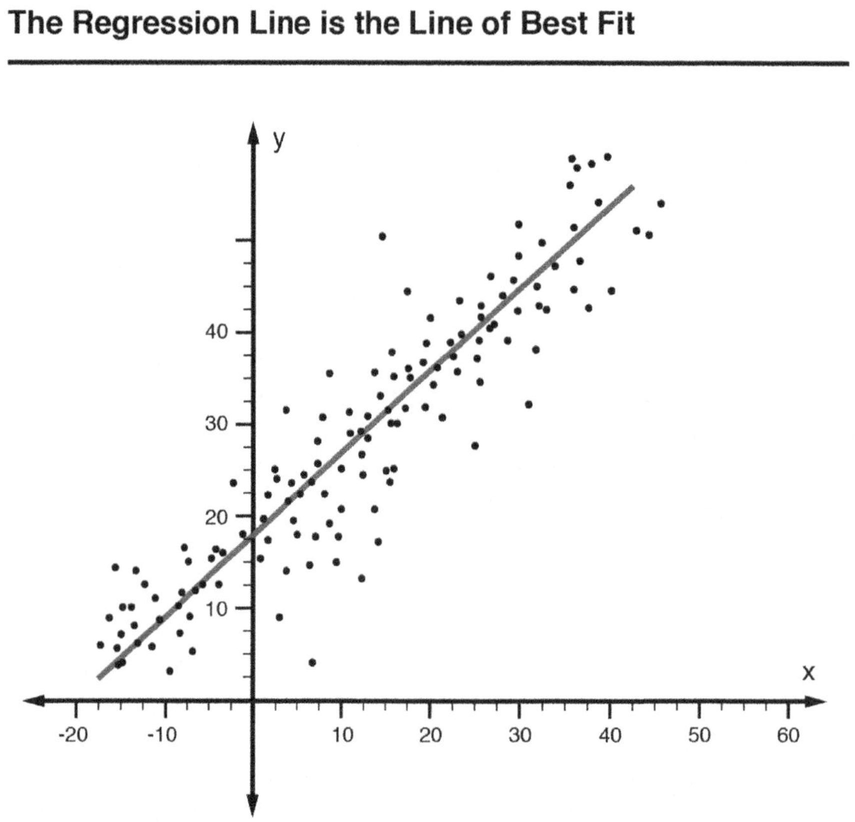

Regression models are highly used for forecasting, and linear regression techniques are the simplest models. If the nonlinear data follows the shape of exponential, logarithmic, or power functions, those types of functions can be used to more accurately model the data rather than lines.

Here is an example of both an exponential regression and a logarithmic regression model:

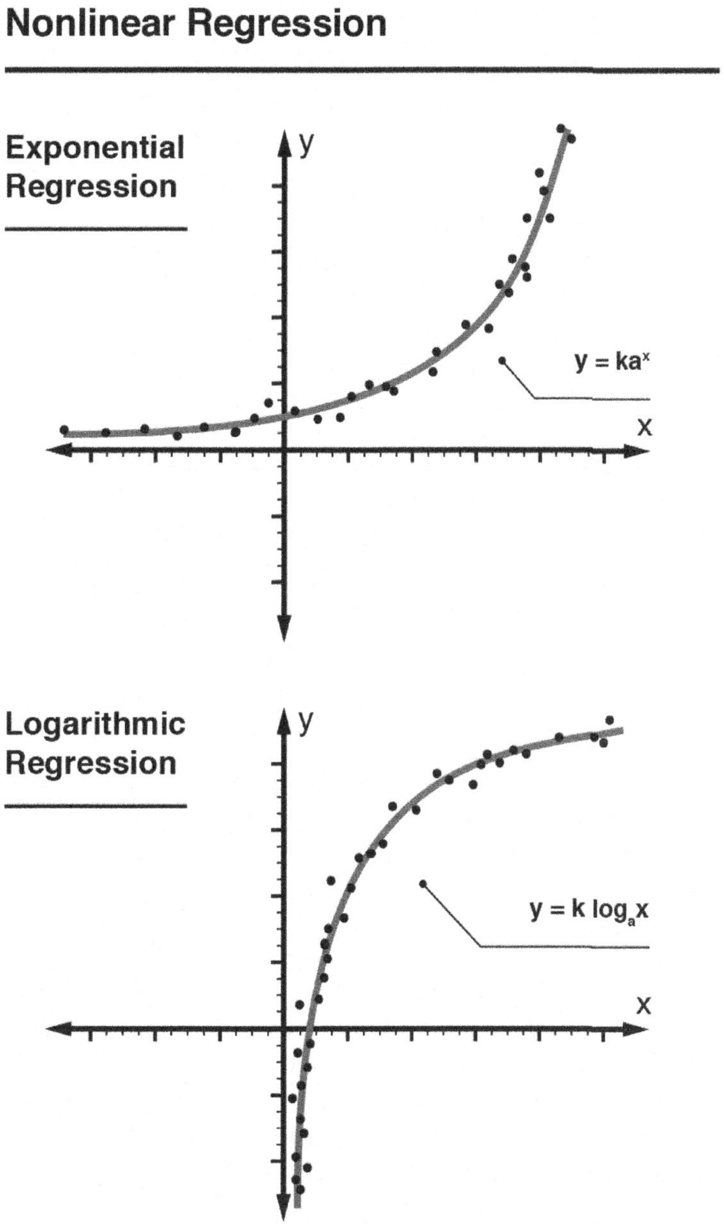

Regression lines can be used to estimate data points not already given. For example, if an equation of a line is found that fit the temperature and beach visitor data set, its input is the average daily temperature and its output is the projected number of visitors. Thus, the number of beach visitors on a 100-degree day can be estimated. The output is a data point on the regression line, and the number of daily visitors is expected to be greater than on a 96-degree day because the regression line has a positive slope.

Once the function is found that fits the data, its accuracy can be calculated. Therefore, how well the line fits the data can be determined. The difference between the actual dependent variable from the data set and the estimated value located on the regression line is known as a **residual.** Therefore, the residual is known as the predicted value $\hat{y}$ minus the actual value y. A residual is calculated for each data point and can be plotted on the scatterplot. If all the residuals appear to be approximately the same distance from the regression line, the line is a good fit. If the residuals seem to differ greatly across the board, the line isn't a good fit.

The formula for a regression line is $y = mx + b$, where m is the slope and b is the y-intercept. Both the slope and y-intercept are found in the **Method of Least Squares**, which is the process of finding the equation of the line through minimizing residuals. The slope represents the rate of change in y as x gets larger. Therefore, because y is the dependent variable, the slope actually provides the predicted values given the independent variable. The ***y*-intercept** is the predicted value for when the independent variable equals zero. In the temperature example, the y-intercept is the expected number of beach visitors for a very cold average daily temperature of zero degrees.

The **correlation coefficient (r)** measures the association between two variables. Its value is between -1 and 1, where -1 represents a perfect negative linear relationship, 0 represents no relationship, and 1 represents a perfect positive linear relationship. A **negative linear relationship** means that as x values increase, y values decrease. A **positive linear relationship** means that as x values increase, y values increase. The formula for computing the correlation coefficient is (n is the number of data points):

$$r = \frac{n(\sum xy) - (\sum x)(\sum y)}{\sqrt{n(\sum x^2) - (\sum x)^2}\sqrt{n(\sum y^2) - (\Sigma y)^2}}$$

Both Microsoft Excel® and a graphing calculator can evaluate this easily once the data points are entered. A correlation greater than 0.8 or less than -0.8 is classified as "strong" while a correlation between -0.5 and 0.5 is classified as "weak."

Correlation Versus Causation

Correlation and causation have two different meanings. If two values are **correlated**, there is an association between them. However, correlation doesn't necessarily mean that one variable causes the other. **Causation** (or "cause and effect") occurs when one variable causes the other. Average daily temperature and number of beachgoers are correlated and have causation. If the temperature increases, the change in weather causes more people to go to the beach. However, alcoholism and smoking are correlated but don't have causation. The more someone drinks the more likely they are to smoke but drinking alcohol doesn't cause someone to smoke.

Using Equations of Linear Models to Solve Problems

The slope of the graph of a line represents the rate of change between the variables of an equation. In the context of a real-world scenario, the slope will tell the way in which the unknown quantities (variables) change with respect to each other. A scenario involving distance and time might state that someone is traveling at a rate of 45 miles per hour. The slope of the linear model would be 45. A scenario involving the cost of a cab ride and distance traveled might state that the person is charged $3 for each mile. The slope of the linear model would be 3.

The y-intercept of a linear function is the value of y when $x = 0$ (the point where the line intercepts the y-axis on the graph of the equation). It is sometimes helpful to think of this as a "starting point" for a linear function. Suppose for the scenario about the cab ride that the person is told that the cab company charges a flat fee of $5 plus $3 for each mile. Before traveling any distance ($x = 0$), the cost is $5. The *y*-intercept for the linear model would be 5.

A linear equation with two variables can be written given a point (ordered pair) and the slope or given two points on a line. An ordered pair gives a set of corresponding values for the two variables (*x* and *y*). As an example, for a scenario involving distance and time, it is given that the person traveled 112.5 miles in 2 ½ hours. Knowing that *x* represents time and *y* represents distance, this information can be written as the ordered pair (2.5, 112.5).

Two-Way Frequency Tables

Data that isn't described using numbers is known as **categorical data.** For example, age is numerical data, but hair color is categorical data. Categorical data is summarized using two-way frequency tables. A **two-way frequency table** counts the relationship between two sets of categorical data. There are rows and columns for each category, and each cell represents frequency information that shows the actual data count between each combination. A contingency table presents the frequency tables of both variables simultaneously. The levels of one variable constitute the rows of the table, and the levels of the other constitute the columns. The margins consist of the sum of cell frequencies for each row and each column (marginal frequencies). The lower right corner value is the sum of marginal frequencies for the rows or the sum of the marginal frequencies for the columns. Both are equal to the total sample size.

For example, the graphic on the left-side below is a two-way frequency table showing the number of girls and boys taking language classes in school. Entries in the middle of the table are known as the **joint frequencies**. For example, the number of girls taking French class is 12, which is a joint frequency. The totals are the **marginal frequencies**. For example, the total number of boys is 20, which is a marginal frequency. If the frequencies are changed into percentages based on totals, the table is known as a **two-way relative frequency table.** Percentages can be calculated using the table total, the row totals, or the column totals.

Here's the process of obtaining the two-way relative frequency table using the table total:

Two-Way Frequency Table

	Spanish	French	German	**Total**
Boys	10	2	8	**20**
Girls	15	12	3	**30**
Total	**25**	**14**	**11**	**50**

Joint frequencies

Marginal frequencies

Divide all tables entries by the table total (50) →

Two-Way Relative Frequency Table (with respect to table total)

	Spanish	French	German	**Total**
Boys	0.2	0.04	0.16	**0.40**
Girls	0.3	0.24	0.06	**0.60**
Total	**0.5**	**0.28**	**0.22**	**1.00**

The middle entries are known as **joint probabilities** and the totals are **marginal probabilities.** In this data set, it appears that more girls than boys take Spanish class. However, that might not be the case because more girls than boys were surveyed and the results might be misleading. To avoid such errors, **conditional relative frequencies** are used. The relative frequencies are calculated based on a row or column.

Here are the conditional relative frequencies using column totals:

Two-Way Frequency Table

	Spanish	French	German	**Total**
Boys	10	2	8	**20**
Girls	15	12	3	**30**
Total	**25**	**14**	**11**	**50**

Divide each column entry by that column's total →

Two-Way Relative Frequency Table (with respect to table total)

	Spanish	French	German	**Total**
Boys	0.4	0.14	0.73	**0.4**
Girls	0.6	0.86	0.27	**0.6**
Total	**1.00**	**1.00**	**1.00**	**1.00**

Two-way frequency tables can help in making many conclusions about the data. If either the row or column of conditional relative frequencies differs between each row or column of the table, then an association exists between the two categories. For example, in the above tables, the majority of boys are taking German while the majority of girls are taking French. If the frequencies are equal across the rows, there is no association and the variables are labelled as independent. It's important to note that the association does exist in the above scenario, though these results may not occur the next semester when students are surveyed.

When measuring event probabilities, two-way frequency tables can be used to report the raw data and then used to calculate probabilities. If the frequency tables are translated into relative frequency tables, the probabilities presented in the table can be plugged directly into the formulas for conditional probabilities. By plugging in the correct frequencies, the data from the table can be used to determine if events are independent or dependent.

Practice Questions

1. Karen gets paid a weekly salary and a commission for every sale that she makes. The table below shows the number of sales and her pay for different weeks.

Sales	2	7	4	8
Pay	\$380	\$580	\$460	\$620

Which of the following equations represents Karen's weekly pay?

a. $y = 90x + 200$
b. $y = 90x - 200$
c. $y = 40x + 300$
d. $y = 40x - 300$

2. The phone bill is calculated each month using the equation $c = 50g + 75$. The cost of the phone bill per month is represented by c, and g represents the gigabytes of data used that month. What is the value and interpretation of the slope of this equation?

a. 75 dollars per day
b. 75 gigabytes per day
c. 50 dollars per day
d. 50 dollars per gigabyte

3. Arrange the following numbers from least to greatest value:

$0.85, \frac{4}{5}, \frac{2}{3}, \frac{91}{100}$

a. $0.85, \frac{4}{5}, \frac{2}{3}, \frac{91}{100}$
b. $\frac{4}{5}, 0.85, \frac{91}{100}, \frac{2}{3}$
c. $\frac{2}{3}, \frac{4}{5}, 0.85, \frac{91}{100}$
d. $0.85, \frac{91}{100}, \frac{4}{5}, \frac{2}{3}$

4. Which graph will be a line parallel to the graph of $y = 3x - 2$?

a. $2y - 6x = 2$
b. $y - 4x = 4$
c. $3y = x - 2$
d. $2x - 2y = 2$

5. An equation for the line passing through the origin and the point $(2, 1)$ is

a. $y = 2x$
b. $y = \frac{1}{2}x$
c. $y = x - 2$
d. $2y = x + 1$

6. A line goes through the point (-4, 0) and the point (0, 2). What is the slope of the line?

a. 2

b. 4

c. $\frac{3}{2}$

d. $\frac{1}{2}$

7. What are the y-intercept(s) for $y = x^2 + 3x - 4$?

a. $y = 1$
b. $y = -4$
c. $y = 3$
d. $y = 4$

8. Write the expression for six less than three times the sum of twice a number and one.

a. $2x + 1 - 6$
b. $3x + 1 - 6$
c. $3(x + 1) - 6$
d. $3(2x + 1) - 6$

9. What is the volume of a cylinder, in terms of π, with a radius of 5 inches and a height of 10 inches?

a. 250 π in^3
b. 50 π in^3
c. 100 π in^3
d. 200 π in^3

10. What is the volume of a cylinder, in terms of π, with a radius of 6 centimeters and a height of 2 centimeters?

a. 36 π cm^3
b. 24 π cm^3
c. 72 π cm^3
d. 48 π cm^3

11. What would be the coordinates if the point plotted on the grid was reflected over the x-axis?

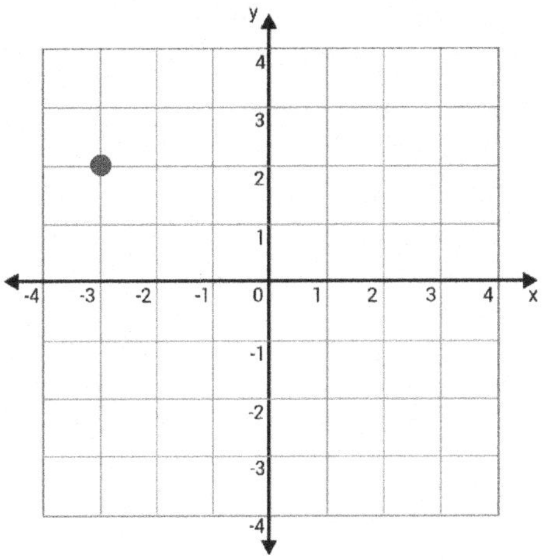

a. (-3, 2)
b. (2, -3)
c. (-3, -2)
d. (2, 3)

12. Katie works at a clothing company and sold 192 shirts over the weekend. $^1/_3$ of the shirts that were sold were patterned, and the rest were solid. Which mathematical expression would calculate the number of solid shirts Katie sold over the weekend?

a. $192 \times \frac{1}{3}$

b. $192 \div \frac{1}{3}$

c. $192 \times (1 - \frac{1}{3})$

d. $192 \div 3$

13. For the following similar triangles, what are the values of x and y (rounded to one decimal place)?

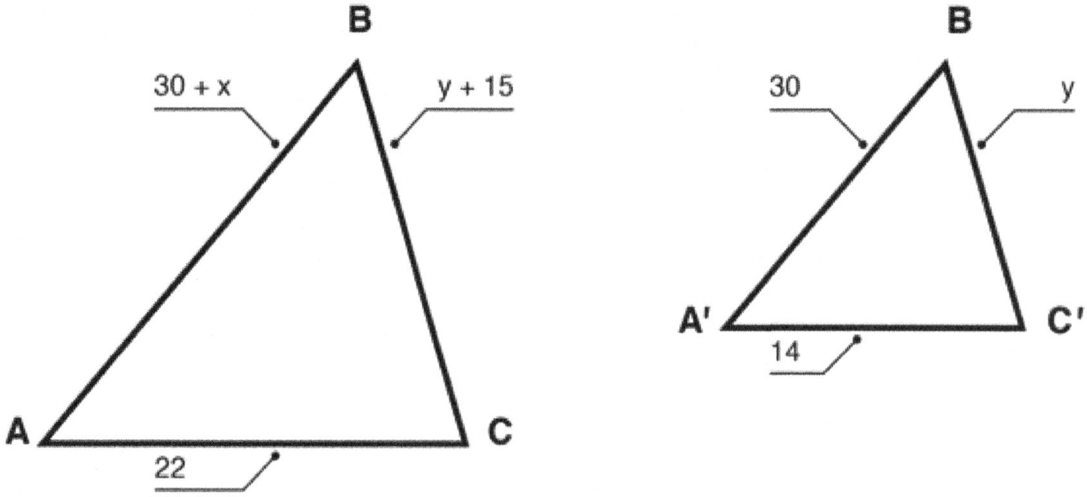

a. $x = 16.5, y = 25.1$
b. $x = 19.5, y = 24.1$
c. $x = 17.1, y = 26.3$
d. $x = 26.3, y = 17.1$

14. If the volume of a sphere is 288π cubic meters, what are the radius and surface area of the same sphere?
a. Radius 6 meters and surface area 144π square meters
b. Radius 36 meters and surface area 144π square meters
c. Radius 6 meters and surface area 12π square meters
d. Radius 36 meters and surface area 12π square meters

15. What is the length of the hypotenuse of a right triangle with one leg equal to 3 centimeters and the other leg equal to 4 centimeters?
a. 7 cm
b. 5 cm
c. 25 cm
d. 12 cm

16. What is the solution for the following equation?

$$\frac{x^2 + x - 30}{x - 5} = 11$$

a. $x = -6$
b. There is no solution.
c. $x = 16$
d. $x = 5$

17. What's the distance between the points $(-1, 2)$ and $(3, -6)$?
a. 5.66
b. 32
c. 80
d. 8.94

18. What is the slope of this line?

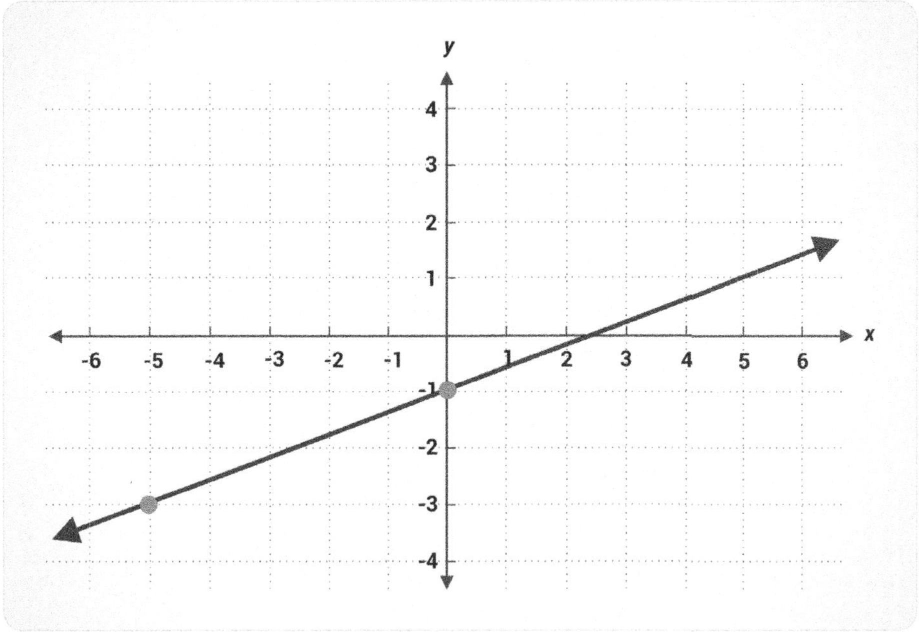

a. 2

b. $\frac{5}{2}$

c. $\frac{1}{2}$

d. $\frac{2}{5}$

19. Which of the following equations best represents the problem below?

The width of a rectangle is 2 centimeters less than the length. If the perimeter of the rectangle is 44 centimeters, then what are the dimensions of the rectangle?

a. $2l + 2(l - 2) = 44$
b. $l + 2) + (l + 2) + l = 48$
c. $l \times (l - 2) = 44$
d. $(l + 2) + (l + 2) + l = 44$

20. In the image below, what is demonstrated by the two triangles?

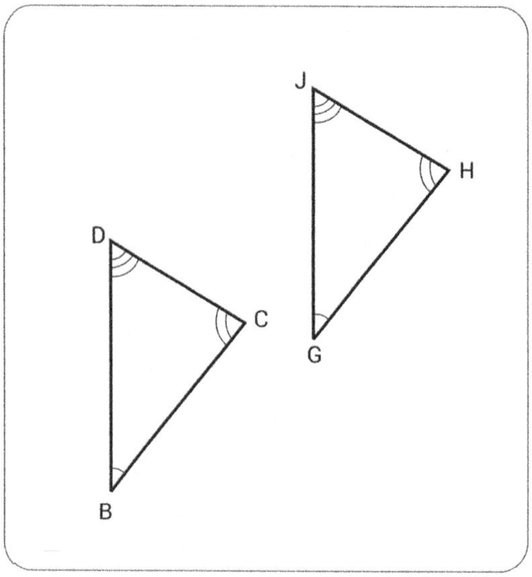

a. According to Side-Side-Side, the triangles are congruent.
b. According to Angle-Angle-Angle, the triangles are congruent.
c. According to Angle-Side-Angle, the triangles are congruent.
d. There is not enough information to prove the two triangles are congruent.

21. In the image below, what is demonstrated by the two triangles?

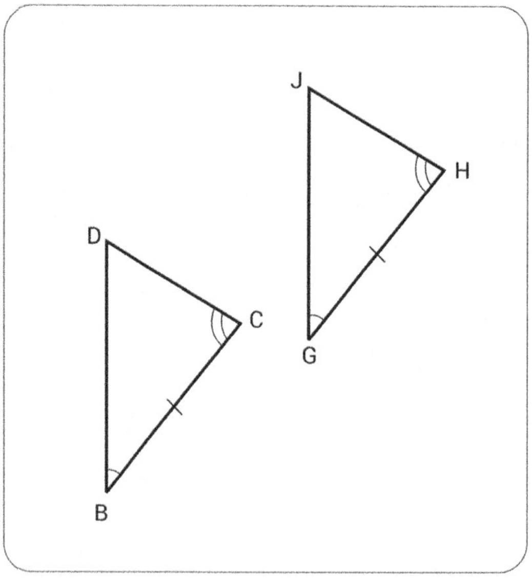

a. According to Side-Side-Side, the triangles are congruent.
b. According to Angle-Angle-Angle, the triangles are congruent.
c. According to Angle-Side-Angle, the triangles are congruent.
d. There is not enough information to prove the two triangles are congruent.

22. In the image below, what is demonstrated by the two triangles?

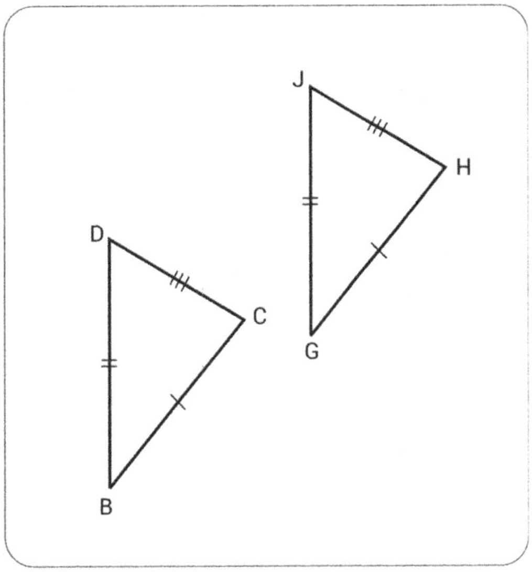

a. According to Side-Side-Side, the triangles are congruent.
b. According to Angle-Angle-Angle, the triangles are congruent.
c. According to Angle-Side-Angle, the triangles are congruent.
d. There is not enough information to prove the two triangles are congruent.

23. What type of function is modeled by the values in the following table?

X	f(x)
1	2
2	4
3	8
4	16
5	32

a. Linear
b. Exponential
c. Quadratic
d. Cubic

24. What is the solution to the following system of equations?

$$x^2 - 2x + y = 8$$
$$x - y = -2$$

a. $(-2, 3)$
b. There is no solution.
c. $(-2, 0)\ (1, 3)$
d. $(-2, 0)\ (3, 5)$

25. If the ordered pair $(-3, -4)$ is reflected over the *x*-axis, what is the new ordered pair?
a. $(-3, -4)$
b. $(3, -4)$
c. $(3, 4)$
d. $(-3, 4)$

26. The variable *y* is directly proportional to *x*. If $y = 3$ when $x = 5$, then what is *y* when $x = 20$?
a. 10
b. 12
c. 14
d. 16

27. What is the solution to the following system of equations?

$$\begin{cases} x^2 + y = 4 \\ 2x + y = 1 \end{cases}$$

a. (−1, 3)
b. (−1, 3), (3, −5)
c. (3, −5)
d. (−1, -5)

28. A company invests $50,000 in a building where they can produce saws. If the cost of producing one saw is $40, then which function expresses the amount of money the company pays? The variable *y* is the money paid and *x* is the number of saws produced.
a. $y = 50{,}000x + 40$
b. $y + 40 = x - 50{,}000$
c. $y = 40x - 50{,}000$
d. $y = 40x + 50{,}000$

29. Which of the following correctly displays 8,600,000,000,000 in scientific notation (to two significant figures)?
a. 8.6×10^{12}
b. 8.6×10^{-12}
c. 8.6×10^{11}
d. 86×10^{12}

30. Which of the following is the best description of the relationship between X and Y?
a. The data has normal distribution.
b. X and Y have a negative relationship.
c. No relationship
d. X and Y have a positive relationship.

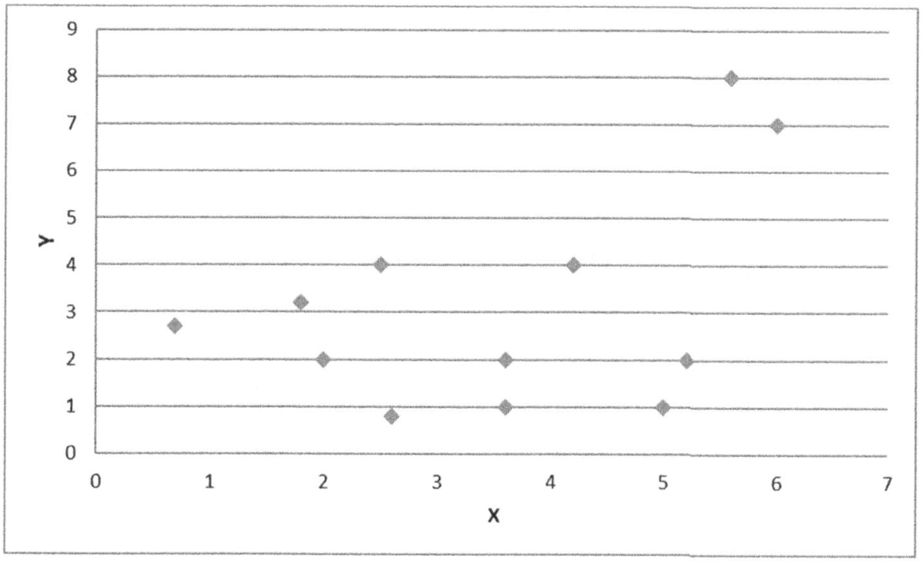

31. Consider a four-year private institution that you would like to attend for college. The tuition per year is $22,000 and it is estimated that room and board will be $5,000 per year and books will cost you $500 per year. If your family is expected to pay for 25% of your total expenses, how much will you have to have in a savings account if you would like to have the total amount available before you even attend?
a. $82,500
b. $110,000
c. $88,000
d. $27,500

32. If $-3(x+4) \geq x+8$, what is the value of x?
a. $x = 4$
b. $x \geq 2$
c. $x \geq -5$
d. $x \leq -5$

33. What is the scale factor of the following dilation where square ABCD represents the original shape and A'B'C'D' represents the dilation?

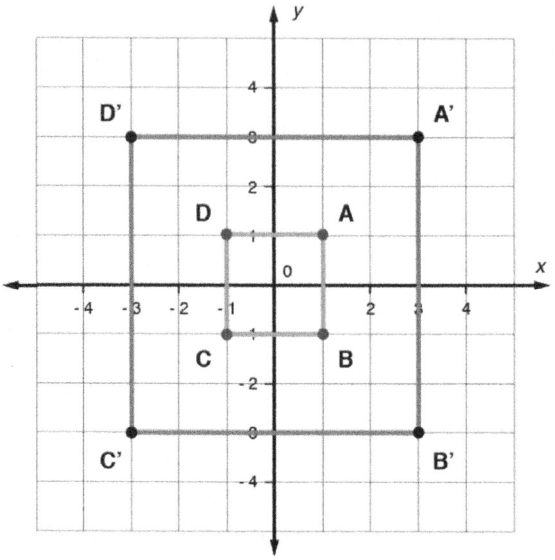

a. 2
b. 1
c. 4
d. 3

34. If a car can travel 300 miles in 4 hours, how far can it go in an hour and a half?
a. 100 miles
b. 112.5 miles
c. 135.5 miles
d. 150 miles

35. At the store, Jan spends $90 on apples and oranges. Apples cost $1 each and oranges cost $2 each. If Jan buys the same number of apples as oranges, how many oranges did she buy?
a. 20
b. 25
c. 30
d. 35

36. Which of the following statements is true about the two lines below?

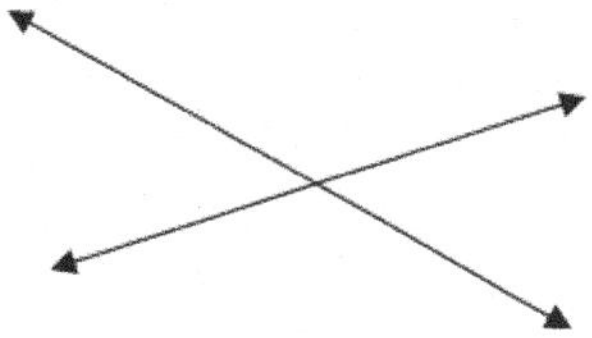

a. The two lines are parallel but not perpendicular.
b. The two lines are perpendicular but not parallel.
c. The two lines are both parallel and perpendicular.
d. The two lines are neither parallel nor perpendicular.

37. What's the midpoint of a line segment with endpoints $(-1, 2)$ and $(3, -6)$?

a. $(1, 2)$
b. $(1, 0)$
c. $(-1, 2)$
d. $(1, -2)$

38. It costs Shea $12 to produce 3 necklaces. If he can sell each necklace for $20, how much profit would he make if he sold 60 necklaces?

a. $240
b. $360
c. $960
d. $1200

39. What is the sum of $(2.6 \times 10^5) + (1.3 \times 10^4)$? Enter your answer in the answer box below.

40. The area of a given rectangle is 24 square centimeters. If the measure of each side is multiplied by 3, what is the area of the new figure? Enter your answer in the box below.

41. If $6t + 4 = 16$, what is t?

42. What is the value of x for the right triangle shown below?

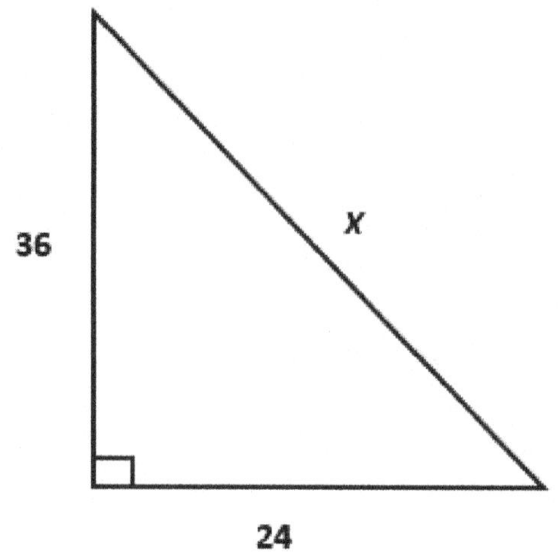

43. A train traveling 50 miles per hour takes a trip lasting 3 hours. If a map has a scale of 1 inch per 10 miles, how many inches apart are the train's starting point and ending point on the map?

44. Bernard can make $80 per day. If he needs to make $300 and only works full days, how many days will this take?

45. What is the length of the other leg of a right triangle with a hypotenuse of 10 inches and a leg of 8 inches?

46. If Sarah reads at an average rate of 21 pages in four nights, how many nights will it take her to read 140 pages?

47. A truck is carrying three cylindrical barrels. Their bases have a diameter of 2 feet, and they have a height of 3 feet. In terms of π, what is the total volume of the three barrels in cubic feet?

48. Jessica buys 10 cans of paint. Red paint costs $1 per can and blue paint costs $2 per can. In total, she spends $16. How many red cans did she buy?

49. Apples cost $2 each, while bananas cost $3 each. Maria purchased 10 fruits in total and spent $22. How many apples did she buy?

50. What is the volume of a cone, in terms of π, with a radius of 10 centimeters and height of 12 centimeters?

Answer Explanations

1. C: $y = 40x + 300$. In this scenario, the variables are the number of sales and Karen's weekly pay. The weekly pay depends on the number of sales. Therefore, weekly pay is the dependent variable (y), and the number of sales is the independent variable (*x*). Each pair of values from the table can be written as an ordered pair (*x*, *y*): (2, 380), (7, 580), (4, 460), (8, 620). The ordered pairs can be substituted into the equations to see which creates true statements (both sides equal) for each pair. Even if one ordered pair produces equal values for a given equation, the other three ordered pairs must be checked. The only equation which is true for all four ordered pairs is $y = 40x + 300$:

$$380 = 40(2) + 300 \rightarrow 380 = 380$$

$$580 = 40(7) + 300 \rightarrow 580 = 580$$

$$460 = 40(4) + 300 \rightarrow 460 = 460$$

$$620 = 40(8) + 300 \rightarrow 620 = 620$$

2. D: The slope from this equation is 50, and it is interpreted as the cost per gigabyte used. Since the *g*-value represents number of gigabytes and the equation is set equal to the cost in dollars, the slope relates these two values. For every gigabyte used on the phone, the bill goes up 50 dollars.

3. C: The first step is to depict each number using decimals. $\frac{91}{100}$ = 0.91

Dividing the numerator by denominator of $\frac{4}{5}$ to convert it to a decimal yields 0.80, while $\frac{2}{3}$ becomes 0.66 recurring. Rearrange each expression in ascending order, as found in answer *C*.

4. A: Parallel lines have the same slope. The slope of *C* can be seen to be 1/3 by dividing both sides by 3. The others are in standard form $Ax + By = C$, for which the slope is given by $\frac{-A}{B}$. The slope of A is 3; the slope of B is 4. The slope of *D* is 1.

5. B: The slope will be given by $\frac{1-0}{2-0} = \frac{1}{2}$. The *y*-intercept will be 0, since it passes through the origin. Using slope-intercept form, the equation for this line is $y = \frac{1}{2}x$.

6. D: The slope is given by the change in *y* divided by the change in *x*. The change in *y* is 2-0 = 2, and the change in *x* is 0 – (-4) = 4. The slope is $\frac{2}{4} = \frac{1}{2}$.

7. B: The y-intercept of an equation is found where the x -value is zero. Plugging zero into the equation for x, the first two terms cancel out, leaving -4.

8. D: The expression is three times the sum of twice a number and 1, which is $3(2x + 1)$. Then, 6 is subtracted from this expression.

9. A: The volume of a cylinder is $\pi r^2 h$, and $\pi \times 5^2 \times 10$ is $250\ \pi\ in^3$. Choice *B* is not the correct answer because that is $5^2 \times 2\pi$. Choice *C* is not the correct answer since that is $5in \times 10\pi$. Choice *D* is not the correct answer because that is $10^2 \times 2in$.

10. C: The volume of a cylinder is $\pi r^2 h$, and $\pi \times 6^2 \times 2$ is $72\,\pi$ cm³. Choice *A* is not the correct answer because that is only $6^2 \times \pi$. Choice *B* is not the correct answer because that is $2^2 \times 6 \times \pi$. Choice *D* is not the correct answer because that is $2^3 \times 6 \times \pi$.

11. C: (-3, -2). The coordinates of a point are written as an ordered pair (*x*, *y*). To determine the *x*-coordinate, a line is traced directly above or below the point until reaching the *x*-axis. This step notes the value on the *x*-axis. In this case, the *x*-coordinate is -3. To determine the *y*-coordinate, a line is traced directly to the right or left of the point until reaching the y-axis, which notes the value on the *y*-axis. In this case, the *y*-coordinate is 2. Therefore, the current ordered pair is written (-3, 2). To reflect the point over the *x*-axis, the *x*-coordinate stays the same, but the *y*-coordinate becomes -2, since after a refection over the *x*-axis, the point would be located equivalently far below the *x*-axis as it is above the *x*-axis. Since the current *y*-coordinate is 2, the *y*-coordinate would become -2.

12. C: $\frac{1}{3}$ of the shirts sold were patterned. Therefore, $1 - \frac{1}{3} = \frac{2}{3}$ of the shirts sold were solid. Anytime "of" a quantity appears in a word problem, multiplication should be used. Therefore:

$$192 \times \frac{2}{3} = \frac{192 \times 2}{3} = \frac{384}{3} = 128 \text{ solid shirts were sold}$$

The entire expression is $192 \times \left(1 - \frac{1}{3}\right)$.

13. C: Because the triangles are similar, the lengths of the corresponding sides are proportional. Therefore:

$$\frac{30 + x}{30} = \frac{22}{14} = \frac{y + 15}{y}$$

This results in the equation $14(30 + x) = 22 \cdot 30$ which, when solved, gives $x = 17.1$. The proportion also results in the equation $14(y + 15) = 22y$ which, when solved, gives $y = 26.3$.

14. A: Because the volume of the given sphere is 288π cubic meters, this means ${}^4/_3\,\pi r^3 = 288\pi$. This equation is solved for r to obtain a radius of 6 meters. The formula for the surface area of a sphere is $4\pi r^2$, so if $r = 6$ in this formula, the surface area is 144π square meters.

15. B: This answer is correct because $3^2 + 4^2$ is $9 + 16$, which is 25. Taking the square root of 25 is 5. Choice *A* is not the correct answer because that is $3 + 4$. Choice *C* is not the correct answer because that is stopping at $3^2 + 4^2$ is $9 + 16$, which is 25. Choice *D* is not the correct answer because that is 3×4.

16. B: The equation can be solved by factoring the numerator into $(x + 6)(x - 5)$. Since that same factor $(x - 5)$ exists on top and bottom, that factor cancels. This leaves the equation $x + 6 = 11$. Solving the equation gives the answer $x = 5$. When this value is plugged into the equation, it yields a zero in the denominator of the fraction. Since this is undefined, there is no solution.

17. D: The distance formula is derived from the Pythagorean theorem. To calculate the distance between two coordinates, the following formula is used:

$$d = \sqrt{(x_2 - x_1)^2 + (y_2 - y_1)^2}$$

Plugging in the coordinates $(-1, 2)$ and $(3, -6)$ yields:

$$d = \sqrt{[3-(-1)]^2 + (-6-2)^2}$$

$$d = \sqrt{(4)^2 + (-8)^2}$$

$$d = \sqrt{80} = 8.94$$

18. D: The slope is given by the change in *y* divided by the change in *x*. Specifically, it's:

$$slope = \frac{y_2 - y_1}{x_2 - x_1}$$

The first point is (-5, -3) and the second point is (0, -1). Work from left to right when identifying coordinates. Thus, the point on the left is point 1 (-5, -3) and the point on the right is point 2 (0, -1).

Now we need to just plug those numbers into the equation:

$$slope = \frac{-1-(-3)}{0-(-5)}$$

It can be simplified to:

$$slope = \frac{-1+3}{0+5}$$

$$slope = \frac{2}{5}$$

19. A: The first step is to determine the unknown, which is in terms of the length, l.

The second step is to translate the problem into the equation using the perimeter of a rectangle, $P = 2l + 2w$. The width is the length minus 2 centimeters. The resulting equation is $2l + 2(l-2) = 44$. The equation can be solved as follows:

$2l + 2l - 4 = 44$	Apply the distributive property on the left side of the equation
$4l - 4 = 44$	Combine like terms on the left side of the equation
$4l = 48$	Add 4 to both sides of the equation
$l = 12$	Divide both sides of the equation by 4

The length of the rectangle is 12 centimeters. The width is the length minus 2 centimeters, which is 10 centimeters. Checking the answers for length and width forms the following equation:

$$44 = 2(12) + 2(10)$$

The equation can be solved using the order of operations to form a true statement: $44 = 44$.

20. D: AAA does not prove congruence, only similarity. Choice *A* is not the correct answer because three congruent sides are not shown. Choice *B* is not the correct answer because AAA does not prove congruence, only similarity. Choice *C* is not the correct answer because Angle-Side-Angle is not shown.

21. C: The picture demonstrates Angle-Side-Angle congruence. Choice *A* is not the correct answer because the picture does not show Side-Side-Side congruence. Choice *B* is not the correct answer because Angle-Angle-Angle does not even prove congruence. Choice *D* is not the correct answer because there is enough information to prove congruence.

22. A: The three sides are shown to be congruent. Choice *B* is not the correct answer because Angle-Angle-Angle does not prove congruence. Choice *C* is not the correct answer because Angle-Side-Angle is not shown. Choice D is not the correct answer because there is enough information to prove the triangles are congruent, since all three sides are congruent (SSS).

23. B: The table shows values that are increasing exponentially. The differences between the inputs are the same, while the differences in the outputs are changing by a factor of 2. The values in the table can be modeled by the equation $f(x) = 2^x$.

24. D: This system of equations involves one quadratic function and one linear function, as seen from the degree of each equation. One way to solve this is through substitution. Solving for y in the second equation yields $y = x + 2$. Plugging this equation in for the y of the quadratic equation yields:

$$x^2 - 2x + x + 2 = 8$$

Simplifying the equation, it becomes $x^2 - x + 2 = 8$. Setting this equal to zero and factoring, it becomes:

$$x^2 - x - 6 = 0 = (x - 3)(x + 2)$$

Solving these two factors for x gives the zeros $x = 3, -2$. To find the y-value for the point, each number can be plugged in to either original equation. Solving each one for y yields the points $(3, 5)$ and $(-2, 0)$.

25. D: When an ordered pair is reflected over an axis, the sign of one of the coordinates must change. When it's reflected over the x-axis, the sign of the x coordinate must change. The *y* value remains the same. Therefore, the new ordered pair is $(-3, 4)$.

26. B: To be directly proportional means that $y = mx$. If *x* is changed from 5 to 20, the value of *x* is multiplied by 4. Applying the same rule to the y-value, also multiply the value of *y* by 4. Therefore, *y* = 12.

27. B: The system can be solved using substitution. Solve the second equation for *y*, resulting in:

$$y = 1 - 2x$$

Plugging this into the first equation results in the quadratic equation:

$$x^2 - 2x + 1 = 4$$

In standard form, this equation is equivalent to $x^2 - 2x - 3 = 0$ and in factored form is:

$$(x - 3)(x + 1) = 0$$

Its solutions are $x = 3$ and $x = -1$. Plugging these values into the second equation results in $y = -5$ and $y = 3$, respectively. Therefore, the solutions are the ordered pairs $(-1, 3)$ and $(3, -5)$.

28. D: For manufacturing costs, there is a linear relationship between the cost to the company and the number produced, with a *y*-intercept given by the base cost of acquiring the means of production, and a slope given by the cost to produce one unit. In this case, that base cost is $50,000, while the cost per unit is $40. So, $y = 40x + 50{,}000$.

29. A: The decimal point for this value is located after the final zero. Because the decimal is moved 12 places to the left in order to get it between the *8* and the *6*, then the resulting exponent is positive, so Choice *A* is the correct answer. Choice *B* is incorrect because the decimal has been moved in the wrong direction. Choice *C* is incorrect because the decimal has been moved an incorrect number of times. Choice *D* is incorrect because numbers in scientific notation are written just with the ones unit followed by the decimal point with the appropriate number of significant figures followed by an indication of the power of ten.

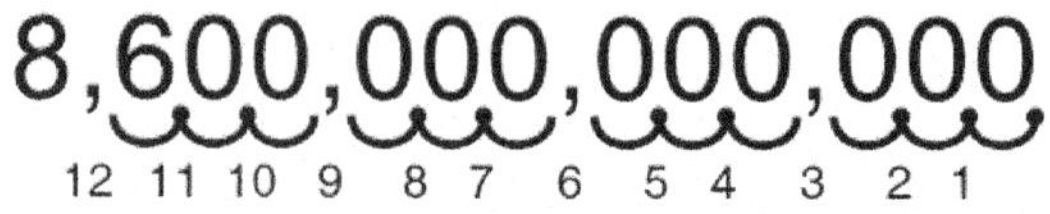

30. C: There is no verifiable relationship between the two variables. While it may seem to have somewhat of a positive correlation because of the last two data points: (5.6, 8) and (6, 7), you must also take into account the two data points before those (5, 1) and (5.2, 2) that have low Y values despite high X values. Data with a normal distribution (Choice *A*) has an arc to it. This data does not.

31. A: Total tuition for four years is:

$$\$22{,}000 \times 4 = \$88{,}000$$

Total room and board for four years is:

$$\$5{,}000 \times 4 = \$20{,}000$$

Total cost of books for 4 years is $\$500 \times 4 = \$2{,}000$. Therefore, it is estimated that you will spend $110,000 on college. Your family is going to pay 25% of this cost, which is:

$$0.25 \times \$110{,}000 = \$27{,}500$$

Therefore, you will be responsible for:

$$\$110{,}000 - \$27{,}500 = \$82{,}500$$

Therefore, to obtain your goal, you will need to have $82,500 in your account before college starts.

32. D: $x \leq -5$. When solving a linear equation or inequality:

Distribution is performed if necessary:

$$-3(x + 4) \to -3x - 12 \geq x + 8$$

This means that any like terms on the same side of the equation/inequality are combined.

The equation/inequality is manipulated to get the variable on one side. In this case, subtracting *x* from both sides produces $-4x - 12 \geq 8$.

The variable is isolated using inverse operations to undo addition/subtraction. Adding 12 to both sides produces $-4x \geq 20$.

The variable is isolated using inverse operations to undo multiplication/division. Remember if dividing by a negative number, the relationship of the inequality reverses, so the sign is flipped. In this case, dividing by -4 on both sides produces $x \leq -5$.

33. D: Consider point A which has coordinates (1, 1). Its corresponding point in the dilation A' has coordinates (3, 3). Therefore, each coordinate has been multiplied times a factor of 3 to obtain the dilation, so 3 is the scale factor. Notice also that each side length has tripled in length from the original length.

34. B: 300 miles in 4 hours is $300 \div 4 = 75$ miles per hour. In 1.5 hours, the car will go 1.5×75 miles, or 112.5 miles.

35. C: One apple/orange pair costs $3 total. Therefore, Jan bought $90 \div 3 = 30$ total pairs, and hence, she bought 30 oranges.

36. D: The two lines are neither parallel nor perpendicular. Parallel lines will never intersect or meet. Therefore, the lines are not parallel. Perpendicular lines intersect to form a right angle (90°). Although the lines intersect, they do not form a right angle, which is usually indicated with a box at the intersection point. Therefore, the lines are not perpendicular.

37. D: The midpoint formula should be used.

$$M = \left(\frac{x_1 + x_2}{2}, \frac{y_1 + y_2}{2}\right) = \left(\frac{-1 + 3}{2}, \frac{2 + (-6)}{2}\right) = (1, -2)$$

38. C: In order to calculate the profit, we need to create an equation that models the total income minus the cost of the materials. $\$60 \times 20 = \$1{,}200$ total income. $60 \div 3 = 20$ sets of materials. $20 \times \$12 = \$240$ cost of materials. $\$1{,}200 - \$240 = \$960$ profit. Choice *A* is not correct, as it is only the cost of materials. Choice *B* is not correct, as it is a miscalculation. Choice *D* is not correct, as it is the total income from the sale of the necklaces.

39. 2.73×10^5: The exponent of the ten must be the same before any operations are performed on the numbers. So, $(2.6 \times 10^5) + (1.3 \times 10^4)$ cannot be added until one of the exponents on the ten is changed. The 1.3×10^4 can be changed to $0.13 * 10^5$, then the 2.6 and 0.13 can be added. The answer comes out to be 2.73×10^5.

40. 216 cm: Because area is a two-dimensional measurement, the dimensions are multiplied by a scale that is squared to determine the scale of the corresponding areas. The dimensions of the rectangle are multiplied by a scale of 3. Therefore, the area is multiplied by a scale of 3^2 (which is equal to 9):

$$24cm \times 9 = 216cm$$

41. 2: First, subtract 4 from each side. This yields $6t = 12$. Now, divide both sides by 6 to obtain $t = 2$.

42. 43.3: The Pythagorean theorem states that for right triangles $c^2 = a^2 + b^2$, with *c* being the side opposite the 90° angle. Substituting 24 as *a* and 36 as *b,* the equation becomes:

$$c^2 = 24^2 + 36^2$$

$$576 + 1296 = 1872$$

The last step is to square both sides to remove the exponent:

$$c = \sqrt{1872} = 43.3$$

43. 15 inches: First, the train's journey in the real word is $3 \times 50 = 150$ miles. On the map, 1 inch corresponds to 10 miles, so there are $150 \div 10 = 15$ inches on the map.

44. 4 days: 300/80 = 30/8 = 15/4 = 3.75. But Bernard is only working full days, so he will need to work 4 days, since 3 days are not sufficient.

45. 6: This answer is correct because $100 - 64$ is 36, and taking the square root of 36 is 6.

46. 27 nights: This problem can be solved by setting up a proportion involving the given information and the unknown value. The proportion is:

$$\frac{21\, pages}{4\, nights} = \frac{140\, pages}{x\, nights}$$

Solving the proportion by cross-multiplying, the equation becomes $21x = 4 \times 140$, where $x = 26.67$. Since it is not an exact number of nights, the answer is rounded up to 27 nights. Twenty-six nights would not give Sarah enough time.

47. 9π cubic feet: The formula for the volume of a cylinder is $\pi r^2 h$, where *r* is the radius and *h* is the height. The diameter is twice the radius, so these barrels have a radius of 1 foot. That means each barrel has a volume of $\pi \times 1^2 \times 3 = 3\pi$ cubic feet. Since there are three of them, the total is:

$$3 \times 3\pi = 9\pi \text{ cubic feet}$$

48. 4 cans: We are trying to find x, the number of red cans. The equation can be set up like this:

$$x + 2(10 - x) = 16$$

The left x is actually multiplied by \$1, the price per red can. Since we know Jessica bought 10 total cans, $10 - x$ is the number blue cans that she bought. We multiply the number of blue cans by \$2, the price per blue can.

That should all equal \$16, the total amount of money that Jessica spent. Working that out gives us:

$$x + 20 - 2x = 16$$

$$20 - x = 16$$

$$x = 4$$

49. 8: Let *a* be the number of apples and *b* the number of bananas. Then, the total cost is $2a + 3b = 22$, while it also known that $a + b = 10$. Using the knowledge of systems of equations, cancel the *b* variables by multiplying the second equation by -3. This makes the equation:

$$-3a - 3b = -30$$

Adding this to the first equation, the b values cancel to get $-a = -8$, which simplifies to $a = 8$.

50. 400π cm³: The volume of a cone is ($\pi r^2 h$), divided by 3, and ($\pi \times 10^2 \times 12$), divided by 3 is 400π cm^3.

Dear Common Core Math Grade 8 Test Taker,

We would like to start by thanking you for purchasing this study guide for your exam. We hope that we exceeded your expectations.

Our goal in creating this study guide was to cover all of the topics that you will see on the test. We also strove to make our practice questions as similar as possible to what you will encounter on test day. With that being said, if you found something that you feel was not up to your standards, please send us an email and let us know.

We have study guides in a wide variety of fields. If you're interested in one, try searching for it on Amazon or send us an email.

Thanks Again and Happy Testing!
Product Development Team
info@studyguideteam.com

FREE Test Taking Tips DVD Offer

To help us better serve you, we have developed a Test Taking Tips DVD that we would like to give you for FREE. **This DVD covers world-class test taking tips that you can use to be even more successful when you are taking your test.**

All that we ask is that you email us your feedback about your study guide. Please let us know what you thought about it – whether that is good, bad or indifferent.

To get your **FREE Test Taking Tips DVD**, email freedvd@studyguideteam.com with "FREE DVD" in the subject line and the following information in the body of the email:

a. The title of your study guide.

b. Your product rating on a scale of 1-5, with 5 being the highest rating.

c. Your feedback about the study guide. What did you think of it?

d. Your full name and shipping address to send your free DVD.

If you have any questions or concerns, please don't hesitate to contact us at freedvd@studyguideteam.com.

Thanks again!

Made in the USA
Coppell, TX
05 June 2021

56923546R00063